Chemical Insect Attractants
and Repellents

Chemical Insect Attractants and Repellents

By

VINCENT G. DETHIER, A.M., Ph.D.

Professor of Zoology and Entomology, The Ohio State University;
Formerly Entomologist, Inter-Allied Malaria Control Commission,
Gold Coast, B. W. A.

Philadelphia · THE BLAKISTON COMPANY · Toronto

To the memory of
DAVID HUNT LINDER,
late Curator of the Farlow Herbarium
of Harvard University, at whose sug-
gestion this work was undertaken and
whose encouragement and assistance were
a constant source of inspiration

Preface

Until now, the bulk of effort expended in the study of attractants and repellents has expressed itself as a search by the trial and error method for chemicals effective for the control of insect pests. In many instances a lamentable lack of understanding of the basic principles underlying this subject has been exhibited. This has been due in large part to a lack of appreciation of the chemical, physical, physiological, and botanical factors involved and to the inaccessibility of much allied information the importance of which had not heretofore been realized. In short, through this combination of circumstances one could not see the forest for the trees.

The opportunity for remedying the chaotic condition of the entire subject and imparting greater impetus to research in this field lies in a better understanding, or at least awareness, of the chemical, physical, physiological, and botanical background immediately involved and the assembly, interpretation, and evaluation of scattered works in terms of this background. To that extent this text represents a theoretical study. It is neither a compilation of recipes for attractant and repellent substances nor a manual on insect control. Excellent treatments of insecticides, fumigants, traps, baits, spraying, and insect control generally are available. For detailed discussion of those subjects the reader is referred to Peterson's *Manual of Entomological Equipment and Methods*, Sweetman's *The Biological Control of Insects*, Wardle and Buckle's *The Principles of Insect Control*, and Shepard's *The Chemistry and Toxicology of Insecticide*.

Though the study of attractants and repellents is essentially one of stimulation and response, it does not involve a comprehensive treatment of chemoreception. The morphological and physiological aspects are treated adequately in Snodgrass's *The Principles of Insect Morphology*, Wigglesworth's *The Principles of Insect Physiology*, and Fraenkel and Gunn's *The Orientation of Animals*. Nor is the subject so narrowly delineated as would be a study comprehended by the term chemoreception. This book represents an attempt to bridge the borderline between chemoreception and the broader aspects of behavior based upon it. It is an attempt at a comprehensive insight into specific chemoreception as a unit of behavior from which might be derived practical procedures, methods, and ideas for

vii

future investigation. On this account the chemical viewpoint is stressed throughout.

It is my conviction that more lively interest in attractants and repellents will lead to a more thorough understanding of insect behavior and ecology and in certain cases to a fuller understanding of the evolution of certain habits and behavior patterns. Answers to some of the puzzles of host-parasite relationships, food plant preferences, and physiological races are also bound up in this study. New vistas may be opened also in the field of sensory physiology, especially as it relates to quantitative evaluations of sense organs.

The final phases of the writing of this book were undertaken while serving with the Armed Forces overseas. Completion of the work under these circumstances was made possible only through the generous assistance of many friends and associates. Special thanks are due Miss Frances M. Dethier and Mrs. Thalia Jillson through whom liaison with libraries "Stateside" was maintained, and to Captain Gerard Dethier, U.S.M.C., who assisted in the preparation of illustrations. The manuscript was read critically in its entirety by Professor C. T. Brues and the late Dr. D. H. Linder of Harvard University and by Dr. L. E. Chadwick, Chief of Entomology Section, Medical Division, Army Chemical Corps, Edgewood Arsenal, Maryland. Professor William H. Weston, Jr., of Harvard University has been especially kind with his time, suggestions, and advice.

All line figures adapted from other sources have been redrawn and in most cases simplified or otherwise modified. Halftones have been reproduced through the kindness of the authors and publishers of the original sources. The source of each is indicated in the text. Nearly all listed references have been consulted directly. A few, marked with an asterisk, are not available in this country. An effort has been made to include in the list of references every pertinent contribution; however, at the end of Chapter 8, because of the relatively enormous volume of literature dealing with repellents, only a selected bibliography was possible. It is hoped that no major contributions have been overlooked.

A great debt of gratitude is due those who have granted permission to use data accumulated during the war but not yet published in the open literature: to Captain E. G. Hakansson, U.S.N., and Lt. (j.g.) L. A. Jachowski, U.S.N., for permission to quote from reports of the Naval Medical Research Institute, National Naval Medical Center; to Dr. Alexander King, Director of the United Kingdom Scientific Mission, British Commonwealth Scientific Office, and Dr. M. F. Day of the Australian Scientific Liaison Office, for permission to quote from unpublished material of British origin; to Dr. Ralph Davidson for permission to quote from work reported in O.S.R.D. reports; to Dr. Hubert Frings for permission to quote from work published in reports of the Army Chemical Corps; to Professor A. C. Hodson of the University of Minnesota for supplying unpublished data on apple maggot bait traps.

Grateful acknowledgment is made to the Society of the Sigma Xi for a Grant-in-Aid to assist in the completion of the manuscript of this book and to the American Academy of Arts and Sciences for a Permanent Science Fund Grant-in-Aid which made possible a large share of the original research reported in Chapter 3.

V. G. DETHIER

East Bluehill, Maine

Contents

PREFACE vii

FOREWORD xiii

1. INTRODUCTION 1
   History—Definition of Terms—Thermostimuli—Photostimuli—Mechanostimuli.

2. NATURE OF CHEMICAL ATTRACTANTS 12
   Odors and Tastes—Gradient Fields—Humidity Reactions—Source of Odors and Tastes in Nature—Rôle of Odorous Attractants in Nature—Sex Attractants— Chemistry of Sex Scents—Recognition Odors—Ovipositional-type Attractants— Food-type Attractants—Attractants of Doubtful Significance.

3. ESSENTIAL OILS, RESINS, AND RELATED SUBSTANCES. . . . 39
   Food Plant Selection—Source of Attractive Odors—Mustard Oils and Related Substances—Benzaldehyde and Hydrocyanic Acid—Oils from Umbelliferae—Organic Bases—Attractants in Peach—Aromatic Esters—Attractants in Corn—Oxalic Acid—Solanaceae—Lactones—Geraniol—Discussion.

4. FERMENTATION PRODUCTS (ALCOHOLS, ACIDS, ALDEHYDES, ESTERS, CARBINOLS) 75
   Source of Attractants—Types of Insects Responding to These Attractants—Codling Moth Attractants—Oriental Fruit Moth Attractants—Attractants of Cabbage Looper Moths—Fruit Fly Attractants—Fermentation Products Attractive to Xylophagous Insects—Discussion.

5. PROTEIN AND FAT DECOMPOSITION PRODUCTS (FATTY ACIDS, AMINES, AMMONIA, CARBON DIOXIDE) 103
   Source of Attractants—Occurrence of These Attractants in Nature—Skatole, Indole, Mercaptans, Sulfides—Amines—Fatty Acids—Ammonia—Miscellaneous Gases— Attractants of Mycetophagous Insects—Attractants of Phycophagous Insects—Discussion.

6. OLFACTOMETERS AND THRESHOLD CONCENTRATIONS . . . 140
   Measurement of Odors—Y-tube Type Olfactometers—Venturi Type Olfactometers— Methods of Determining Absolute Threshold Concentrations—Molecular Concentration—Conditions Affecting Thresholds of Response—Taste Thresholds.

7. BAITS AND TRAPS 171
   Application of Basic Principles—Uses of Attractants—Poison Baits—Trap Baits— Traps—Color—Position and Location of Traps—Sex Ratio—Reliability of Trapping as a Means of Sampling Population—Efficiency of Trapping as a Means of Control.

8. REPELLENTS 201

Definition and Action—Physical Repellents—Naturally Occurring Chemical Repellents—Olfactory Repellents—Screening and Evaluating Candidate Repellents—Repellents Extracted from Plants—Synthetic Repellents—Plant-feeding Insects—Honey Bees—Mosquitoes and Blood-sucking Flies—Synergistic and Antagonistic Action of Solvents on Repellents—Impregnated Clothing and Fabrics—Ticks and Chiggers—Blowflies—Houseflies—Termites and Other Wood-boring Insects—Subterranean and Root-dwelling Insects—Mothproofing—Miscellaneous—Discussion.

9. CHEMICAL BASIS OF TASTE AND OLFACTION 233

Taste-Substances—Alcohols—Sugars—Acids—Salts—Odors—Modalities—Acceptance and Rejection—Intensity of Odor—Boiling Point—Molecular Weight—Solubility.

10. EVOLUTION OF FEEDING PREFERENCES 249

Mechanism of Choice—Monophagy, Oligophagy, and Polyphagy—Genetic Basis of Selection—Conditioning—Hopkins Host Principle and Biological Races—Evolution of Feeding Habits—Conclusion.

AUTHOR INDEX 265

SUBJECT INDEX 271

Foreword

No complete and reliable account of the way in which various chemicals attract and repel insects has hitherto been available in a form which takes into account the fundamental basis of such behavior. The present volume by Professor Dethier embodies such a treatment and it is with great satisfaction that I have embraced the opportunity to write a brief foreword to his noteworthy contribution.

The matters with which he deals have always been of deep interest to students of insect ecology, and have long intrigued the curiosity of other biologists less minutely acquainted with the peculiarly specialized behavior of the teeming insect world. They have, furthermore, been of immediate and highly practical importance to those engaged in mitigating the evils imposed by insects upon the human species and his worldly possessions, more particularly the benefits derived from his cultivated crops.

When economic entomology first saw the light as a vaguely defined art designed to combat the ravages of insects affecting agriculture, its basis was empirical in the extreme and this condition persisted well into the present century. Only during the past several decades has there developed any consistent movement to deal with such problems on the basis of sound scientific inquiry. This has been furthered by several developments in related fields of biology, in chemistry, and to a lesser degree in physics.

The rapid increase in our knowledge of insect physiology has gone hand in hand with the advance of general experimental physiology and has become a part of the equipment with which entomologists have been able and willing to arm themselves in the war against noxious insects. Insect ecology similarly has come into its own through the sound application of principles and methods previously utterly outside its scope of action.

Coincident with these advances and obviously, at least in part, contributing to their consummation has been the almost revolutionary progress of chemistry in elucidating the structure of innumerable organic compounds and in developing methods for the synthesis of a vast series of substances, many preconceived on theoretical grounds. Through the medium of biological chemists it has been possible to turn these discoveries to the furtherance of more purely biological research, and entomology is sharing richly in the profits.

With this background we have been able to learn much concerning the behavior of insects with reference to chemicals as these occur in their own bodies, in the food they eat, or elsewhere in their immediate environment. New and highly efficacious insecticides have been developed and the varied reactions of insects to substances which repel or attract them have borne fruit of great economic value. Such studies might at first sight appear to be matters of a very simple type and to represent research as it has recently been degraded and sadly misapplied in certain quarters by the lay public. Closer consideration reveals, however, that this is a wholly inadequate view, for it is now evident that most biological phenomena involve many interacting factors and that we cannot at present simplify them until they are broken down into their component parts. These are not manifest on superficial examination, but must be identified and evaluated by experimental methods. Likewise, other apparently inexplicable phenomena may result from systems which are in reality quite simple in action once they are understood. The so-called "botanical sense" of insects in recognizing their food plants falls into this category.

Nowhere are difficulties of this kind more evident than in dealing with the reactions, behavior, and instinctive processes of animals when we attempt to interpret them on a rational physiological basis. Entomology is faced with these problems in a high degree of complexity. On account of the great importance of insects as they affect the health and food supply of the human species, there are forceful economic as well as academic grounds for fostering research to this end. The resulting accomplishments take form slowly through a series of steps, many of which are commonly not integrated at the moment.

The material presented in this volume is an excellent example of a number of varied but related biological phenomena all of which touch directly on a single physiological theme. This is, in brief, the relation of the sense of smell and to a lesser extent, of taste, to the behavior of insects. Like other animals, the insects live in a world of odors, but they are particularly susceptible to them and frequently uncanny in their delicate sensitivity. Indeed, much of their specific behavior is inseparable from the materials in their environment which they recognize through their chemical sense.

The ramifications of this fundamental relationship throughout the field of pure and applied entomology are very complex as will be seen from the text and the extensive citations of pertinent literature that accompany the several chapters. It is most fortunate that this varied material has been made available in such complete and carefully considered form by one whose competence for the task is enhanced by his own extensive contributions.

CHARLES T. BRUES

Harvard University

Chapter 1

Introduction

History. From his beginning man has been prey to the lusts and appetites of hordes of insects. Very early in his history he devised methods of combatting these pests to which he was host. More often than not manual dexterity in the form of slapping and picking, as practiced in true anthropoid fashion, constituted, as it does in large measure to this day, the prime instrument of insect control. Occasionally refinements were incorporated, as for example, the use of spiny leaves from certain Malvaceae by the Kru people of Liberia as instruments for scratching the itch mite. In the course of his ethnical development man learned that some substances applied to the body discouraged insect aggression. Thus originated the idea of repellents. Early repellents consisted in large part of plants or plant products though animal products also were employed. The Bedouins are known to have used camel urine as a hair tonic and medicine. When applied to the face, it left an oily residue which supposedly acted as an insect repellent. In Arabia during ancient times camel drivers, aware of some connection between the presence of ticks and the ill health of their camels, smeared pitch freely on the bodies of their beasts. In India, hemp (*Cannabis indica*) was and still is spread on beds to drive away pestiferous arthropods. Ancient Kru and Bassa peoples in Liberia learned, probably through accident, that smoke drifting up through the framework of their thatched huts not only cured the thatching materials but discouraged diurnal resting of the notorious malaria vector, *Anopheles gambiae*. It is quite true that fewer *A. gambiae* are found in very smoky huts than in others. Crushed leaves, vegetable brews, and expressed oils also were used down through the ages. As was to be expected, the choice and use of repellents became inseparably associated with religious beliefs and tribal customs. Floral distribution and ecology also dictated the plants to be selected. There was little method to this madness, and the choice of efficacious repellents was due for the greater part to chance. The earliest truly chemical repellent in widespread use was Bordeaux Mixture ($CuSO_4 + Ca(OH)_2$) although Pliny earlier still recommended a mixture of red earth and tar to repel ants. It is known also that the Greeks and Romans painted the backs of parchment manuscripts with cedarwood oil to prevent injury by insects.

There is no record of the use of insect attractants by primitive man. Again the earliest report is that of Pliny in which he suggested that a fish be hung adjacent to trees to lure ants away from the foliage. Early lepidopterists recognized the attractive powers that some substances held for insects and exploited this relationship when they initiated the practice of sugaring for moths. The attractiveness of honey to bees and of light to myriads of other insects is proverbial.

Through the years man has been most concerned with simple, quick methods of dissuading insects from inhabiting his dwelling, his raiment, and his person. With the advent of pastoral and agricultural pursuits he was faced with the added problem of preserving his herds and crops from insect depredations. Yet until the end of the nineteenth century no repellents were marketed. The value of being able to *attract* insects was not immediately apparent, with the result that at the present time the preponderance of knowledge bearing on attractants and repellents deals with the latter. In either case early search for knowledge along these lines was motivated solely by the desire to control insect pests. The earliest recorded use of attractants for economic purposes is Coquillett's attempt in 1885 to control grasshoppers in California by means of attractive poisoned baits. Not long afterwards wine growers in Europe originated the practice of erecting traps baited with stale beer, sugared water, old cider, or lees in an effort to control grapevine moths. By 1896 the attracting value of baits of rum, stale beer, or brown sugar was widely recognized by insect collectors. A later development was the improvement of grasshopper baits by the addition of oranges or lemons (Dean, 1914). Still later, in 1922, Parker placed the practice on a chemical basis by substituting amyl alcohol for citrus fruit. Maxwell-Lefroy (1916) had meanwhile perfected an "ideal" fly bait consisting of a 24-hour-old mixture of casein, brown sugar, and water in equal parts. Without a doubt the greatest impetus given in this trend of investigation stemmed from Peterson's early work in 1925 with molasses-yeast baits for peach moths. Within the next few years Yetter (1925), Peterson (1927), Frost (1927), and others tested hundreds of aromatic compounds in a search for one more attractive to orchard insects than their natural food. The search has gone on and continues to go on. With the discovery by Richmond in 1927 of the exceptional powers of attraction geraniol exerts for Japanese beetles, the search for ideal attractants gained additional impetus.

In the meantime, interest in the chemistry of naturally occurring attractants and repellents from a purely scientific point of view had awakened and was slowly gathering momentum. The obstacles in the path of this field of research appeared insurmountable. Barrows (1907) had published the results of his experiments on the responses of *Drosophila* to amyl alcohol, ethyl alcohol, acetic acid, and lactic acid, some of the products of fermentation in bananas. Howlett (1912) had discovered the attractiveness that oil of citronella held for certain species of *Dacus* although he was unable to explain the phenomenon satisfactorily. Nevertheless, the full significance of chemotaxis in the field of entomology remained unappreciated. Curiously enough the pioneer work which directed the way to new endeavors originated not with an entomologist but with a Dutch botanist. In 1910 Verschaffelt published his sole paper dealing with this subject—a short, concise account of experiments with larvae of cabbage butterflies and of a small wasp. Noting that certain plants in the Amsterdam Botanical Gardens were defoliated by larvae of the cabbage butterfly, he sought the reason for such specific attractiveness. Learning that a group of volatile compounds known collectively as mustard oils were common to all the plants attacked, he tested the responses of larvae to these chemicals. The results of his experiments indicated clearly that specific attractants present in plants influenced phytophagous insects in their choice of food. With this seemingly modest beginning the study of naturally occurring attractants and repellents was initiated.

Shortly thereafter the case for further studies in chemotropism, as it was then termed, was championed by Dewitz (1912), Trägardh (1913), Imms (1914), and Hewitt (1917). Interest naturally enough centered around phytophagous insects. McIndoo (1919), by demonstrating the presence of an olfactory sense in caterpillars, lent additional evidence and support to Verschaffelt's statements that plants attracted insects by odors. From this point onward confirmation and expansion of the principles first demonstrated by Verschaffelt accumulated.

It soon became obvious that attractants and repellents were so intimately related that one could not be discussed and understood without inclusion of the other. One is the antithesis of the other yet a single substance may act as both. Two substances differing in concentration alone may elicit diametrically opposed reactions from an insect. It is this interrelationship which renders difficult a simple yet satisfactory definition in either case.

Definition of Terms. Broadly speaking, an arattctant is anything which draws; a repellent, anything which repels. More specifically, any stimulus which elicits a positive *directive* response may be termed an attractant; any stimulus which elicits an avoiding reaction may be termed a repellent. Such responses, believed spontaneous and inherent, normally direct an organism to its proper habitat. As we shall learn later, it is precisely by reason of this aspect of the responses that attractants and repellents lend themselves so readily as agents for the control of insects.

On the basis of a scheme proposed by Fraenkel and Gunn (1940) reactions of this kind constitute what is termed secondary orientation. They may be divided into directed orientation reactions or taxes and undirected locomotory reactions or kineses, the characteristics (e.g., speed) of which depend upon the intensity of stimulation. For a complete discussion of this aspect of the subject, readers are referred to the comprehensive monograph by the authors quoted.

Theoretically, there may exist as many categories of attractants and repellents as there are classes of external stimuli. Thus to begin with, we may recognize two groups of stimuli, chemical and physical. Although this book is intended primarily to encompass the former, the intimate bearing of physical attractants and repellents on the function of chemicals warrants a brief consideration of physical stimuli. These may be divided into the following types: thermo-, photo-, and mechanostimuli, the latter including contact, steady pressure, slowly alternating pressure, and rapidly alternating pressure (sound).

Thermostimuli. Heat up to certain optimum temperatures acts as an attractant and beyond that point as a repellent, but the reactions of insects toward heat do not depend on the degree of temperature alone. For the purpose of present discussion heat may be considered to act by radiation and by convection. Radiant heat acts in intensity and direction. Certain insects are able to orient to radi-

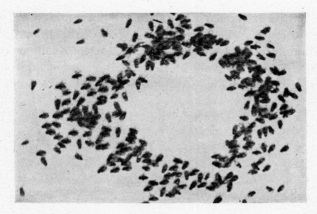

FIG. 1. Flies respond to a temperature gradient by coming to rest in the region of optimum temperature (From Fraenkel and Gunn: "The Orientation of Animals." By permission of Oxford University Press, New York.)

ant heat and may respond to a temperature gradient by arranging themselves at a preferred temperature (Fig. 1). The most numerous and common examples of attraction to heat may be found among those insects parasitic upon warm-blooded animals. Body lice, for example, will migrate toward an artificial finger constructed from a glass tube through which flows warm water (Fig. 2).

Photostimuli. The attractiveness of light to some insects is axiomatic. Less publicized is the attractiveness of darkness to other species. The controversial question of whether or not certain insects

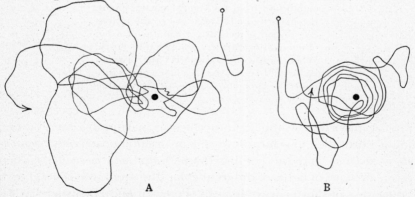

A   B

FIG. 2. The reaction of lice to radiant heat. (A) Artificial finger at temperature of 60.1°–62.5° C. (B) Finger temperature 52.8°–53.3° C. (Redrawn after Homp, *Z. vergleich. Physiol.*, 1938.)

are attracted by darkness (positive scototaxis) or repelled by light (negative phototaxis) lies beyond the scope of this book.

The differential attractiveness of color is an integral part of photo-stimulation. As early as 1888 Plateau demonstrated that insects responded to preferred colors. In nectar and pollen feeders color vision

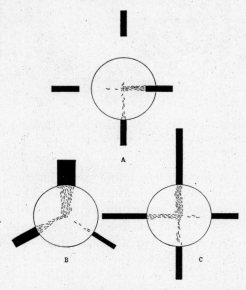

FIG. 3. Diagrams illustrating the reactions of larvae of *Liparis monacha* to dark objects. When objects of equal size are located equidistant from the central starting point, the majority of larvae crawl to the nearest (A). When objects are of equal widths but of unequal heights (C), the tallest is chosen. When widths are varied (B), the broadest is chosen. In each case the chosen object subtends the same visual angle. (Redrawn after Hundertmark, *Z. vergleich. Physiol.*, 1937.)

is most important. Cabbage butterflies are partial to white flowers; sulfur butterflies, to yellow flowers. The oriental peach moth (*Grapholitha molesta*) and the codling moth (*Carpocapsa pomonella*), when given a choice of lights of different color but equal intensities, much prefer rays of shorter wave lengths (Peterson and Haeussler, 1928). Bluish light is decidedly the most attractive. Actually, such responses are not to wave length specifically. Jahn and Crescitelli

(1939) have obtained electrograms illustrating the electrical responses of the dark-adapted moth (*cecropia*) eye to brief exposures of light of different spectral compositions. They found that no specific

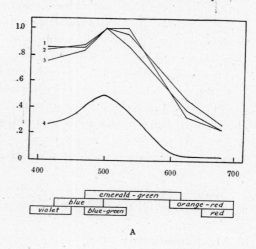

A

Fig. 4. The reactions of insects to different wave lengths of light. (A) Relation between the relative magnitude of the c-wave from electrograms of the cecropia moth eye and wave length in millimicrons. Graphs 1 and 3 are plotted from data obtained with an exposure of 8.3 sigma; graph 2, from data obtained with an exposure of 16.6 sigma. Graph 4 represents the absorption curve of rabbit visual purple. (Redrawn from Jahn and Crescitelli, *J. Cellular Comp. Physiol.*, 1939.)

(B) Group behavior of Coleoptera to 9 wave lengths of light from 436 to 700 millimicrons. Graph 1 represents two tests of 284 *Chrysochus auratus*; graph 2, one test of 79 *Disonycha quinquevittata*; graph 3, two tests of 182 Japanese beetles; graph 4, two tests of 197 Japanese beetles; graph 5, three tests of 216 Japanese beetles. (Redrawn after Weiss, Soraci, and McCoy, *J. N.Y. Entomol. Soc.*, 1943.)

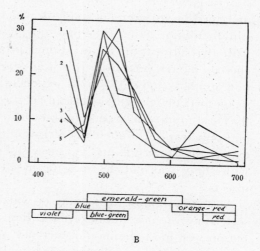

B

effects of wave length on the electrograms were discernible; moreover, if the intensity were properly adjusted, the response to one color could be exactly matched with the response to any other color. This indicates that differences in response to different colors of equal intensity are caused merely by differences in sensitivity and are not effects of wave length *per se*. Weiss (1943, 1943a,* 1944), attacking the problem from the point of view of group behavior, has come to similar conclusions. In short, although wave length *per se* does not appear to affect response, insects react differently to different "colors." Not only the attractiveness of light, therefore, but also the preference for certain colors must be taken into consideration in the design and use of traps.

This point was forcefully brought out by the experiments of Fleming, Burgess, and Maines (1940), and Whittington and Bickley (1941) with Japanese beetles in which it was found that changing the color of traps from yellow-green to yellow raised the total catch from 92.01 to 130.76 pints of beetles. Without a doubt part of the failure of otherwise excellent traps may be laid to an injudicious choice of color.

A color apparently may be equally effective as a repellent in the sense that it is avoided. Ants, for example, consistently avoid ultraviolet.

Mechanostimuli. The importance of omnipresent mechanostimuli is very likely to be overlooked. Hardly a situation can be imagined or a movement made in which some informative contact stimuli are not acting. The most obvious of these arise from contact with the substratum as an animal progresses. Each surface type elicits some response from which an observer may deduce that certain contacts and surfaces are repellent. Among phytophagous insects especially, the physical nature of the substratum may determine the acceptability or nonacceptability of a plant as food. Hairs, spines, and pubescence commonly confer immunity upon plants to such an extent that these plants are successful resistant varieties. Species with shiny wax cuticles may be as effectively repellent to other insects.

Any alteration of the feeding or walking surface to which an insect is accustomed may exert repellent effects for purely physical reasons. Of most common occurrence is dust. Plants bordering unpaved rural roads usually are free from insect attack by virtue of the fine layer of road dusts deposited upon the leaves and flowers. Likewise, artificial or poisonous dusts applied as control measures may act as physical repellents. In both instances particle size is an ex-

---

*An alphabetical code will be used to distinguish the serial order of an author's publications in a single year.

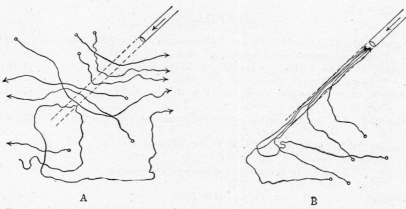

FIG. 5. Diagrams illustrating the response of *Drosophila melanogaster* to a stream of odor-free air (A) and to a stream of air bearing an attractive odor (B). (Redrawn after Flügge, *Z. vergleich. Physiol.*, 1934.)

tremely important factor in determining the repellency of a given dust. This is easily understandable when one considers the mechanical barriers large particles erect to locomotion and feeding by interfering with the proper action of legs and mouthparts.

Water, whether it be in the form of fine bubbles or a film, and sticky materials may act strictly as physical repellents. The latter are widely exploited in the form of tree bands to control gypsy moth larvae, canker worms, and others by preventing their ascension to the crowns of the trees. The antipathy of some insects toward water is convincingly demonstrated with various species of caterpillars. If a ring of water is drawn on a glass plate with a glass rod and a caterpillar placed in the center, the water will act as an efficient barrier. Only after evaporation of the ring will the caterpillar readily crawl beyond its boundaries.

Flying insects ordinarily do not encounter the numerous surface phenomena to which their earth-bound brethren are subject. Both share, however, the mechanical stimulation caused by currents of air and may orient to such currents (anemotaxis). Without orientation to air currents many insects would probably be unable to orient to odors. It appears that initial stimulation is olfactory while subsequent orientation depends upon currents of air (or water) carrying the odor (Fig. 5). This tends to explain why baited traps catch larger numbers of insects when a moderate breeze is blowing than on days when there is not a breath of air.

REFERENCES

Barrows, W. M.: The reactions of the pomace fly, *Drosophila ampelophila* Loew, to odorous substances, *J. Exp. Zoöl.*, **4**(4), 515–537 (1907).

Brandt, H.: Die Lichtorientierung der Mehlmotte *Ephestia kuehniella* Zeller, *Z. vergleich. Physiol.*, **20**(5), 646–673 (1934).

Dean, G. A.: Grasshopper control work in western Kansas, *J. Econ. Entomology*, **7**(1), 67–73 (1914).

Dewitz, J.: The bearing of physiology on economic entomology. *Bull. Entomol. Research*, **3**(4), 343–354 (1912).

Fleming, W. E., E. D. Burgess, and W. W. Maines: The use of traps against the Japanese beetle, *U.S. Dep. Agr. Circ.* 594, 1–11 (1940).

Flügge, C.: Geruchliche Raumorientierung von *Drosophila melanogaster*, *Z. vergleich. Physiol.*, **20**(4), 463–500 (1934).

Fraenkel, G. S., and D. L. Gunn: "The Orientation of Animals," Oxford University Press, 1940.

Frost, S. W.: Further studies of baits for oriental fruit moth control, *J. Econ. Entomology*, **20**(1), 167–174 (1927).

Hewitt, C. G.: Insect behaviour as a factor in applied entomology, *Ibid.*, **10**(1), 81–91 (1917).

Hitti, P. K.: "The Arabs. A Short History," Armed Serv. ed. of "History of the Arabs," Princeton University Press, 1943.

Homp, R.: Wärmeorientierung von *Pediculus vestimenti*, *Z. vergleich. Physiol.*, **26**(1), 1–34 (1938).

Howlett, F. M.: The effect of oil of citronella on two species of Dacus, *Trans. Entomol. Soc. London*, **1912**, 412–418.

Hundertmark, A.: Helligkeits- und Farbenunterscheidungsvermögen der Eiraupen der Nonne (*Lymantria monacha* L.), *Z. vergleich. Physiol.*, **24**(1), 42–57 (1936).

——: Das Formenunterscheidungsvermögen der Eiraupen der Nonne (*Lymantria monacha* L.), *Ibid.*, **24**(4), 563–582 (1937).

Imms. A. D.: The scope and aims of applied entomology, *Parasitology*, **7**(1), 69–87 (1914).

Jahn, T. L., and F. Crescitelli: The electrical responses of the cecropia moth eye. *J. Cellular Comp. Physiol.*, **13**(1), 113–119 (1939).

McIndoo, N. E.: The olfactory sense of lepidopterous larvae, *Ann. Entomol. Soc. Am.*, **12**(2), 65–84 (1919).

——: Smell and taste and their applications, *Sci. Monthly*, **25**, 481–503 (1927).

——: Responses of insects to smell and taste and their value in control, *J. Econ. Entomology*, **21**(6), 903–913 (1928).

——: Tropic responses of codling moth larvae (Abstract), *Ibid.*, **21**(4), 631 (1928a*).

Martini, E.: Zur Kenntnis des Verhaltens der Läuse gegenüber Wärme, *Z. angew. Entomol.*, **4**(1), 34–70 (1917).

Maxwell-Lefroy, H.: A fly destroyer, *Queensland Agr. J.*, **5**(4), 220 (1916).

Parker, J. R.: Improvements in the methods of preparing and using grasshopper baits, *Montana Agr. Exp. Sta. Bull.* 148, 1–19 (1922).

---

* An alphabetical code will be used to distinguish the serial order of an author's publications in a single year.

Peterson, A.: A bait which attracts the oriental peach moth (*Laspeyresia molesta* Busck), *J. Econ. Entomology*, **18**(1), 181–190 (1925).

——: Some baits more attractive to the oriental peach moth than black-strap molasses, *Ibid.*, **20**(1), 174–185 (1927).

——, and G. J. Haeussler: Response of the oriental peach moth and codling moth to colored lights, *Ann. Entomol. Soc. Am.*, **21**(3), 353–379 (1928).

Plateau, F.: Recherches expérimentales sur la vision chez les arthropodes, *Bull. l'acad. roy. Belg.*, Ser. 3, **15**, 28–46 (1888).

Richmond, E. A.: Olfactory response of the Japanese beetle (*Popillia japonica* Newm.), *Proc. Entomol. Soc. Wash.*, **29**(2), 36–44 (1927).

Sen, S. K.: Beginnings in insect physiology and their economic significance, *Agr. J. India.*, **13**(4), 620–627 (1918).

Terherne, R. C.: Bio-chemical aspects of insect control, *Sci. Agr.*, **3**(3), 109–113 (1922).

Thomson, R. C. M.: The reactions of mosquitoes to temperature and humidity, *Bull. Entomol. Research*, **29**(2), 125–140 (1938).

Totze, R.: Beiträge zur Sinnesphysiologie der Zecken, *Z. vergleich. Physiol.*, **19**(1), 110–161 (1933).

Trägardh, I.: On the chemotropism of insects and its significance for economic entomology, *Bull. Entomol. Research*, **4**(2), 113–117 (1913).

Verschaffelt, E.: The cause determining the selection of food in some herbivorous insects, *Proc. Acad. Sci. Amsterdam, Sci. Sec.*, **13**(1), 536–542 (1910).

Wardle, R. A., and P. Buckle: "The Principles of Insect Control," London, Manchester University Press, 1923.

Weiss, H. B.: Color perception in insects, *J. Econ. Entomol.*, **36**(1), 1–17 (1943).

——: Insect behavior to various wave lengths of light, *J. N. Y. Entomol. Soc.*, **51**(2), 117–131 (1943a).

——: Group motor responses of adult and larval forms of insects to different wave lengths of light, *Ibid.*, **52**(1), 27–44 (1944).

Whittington, F. B., and W. E. Bickley: Observations on Japanese beetle traps, *J. Econ. Entomol.*, **34**(2), 219–220 (1941).

Wigglesworth, V. B., and J. D. Gillett: The function of the antennae in *Rhodnius prolixus* (Hemiptera) and the mechanism of orientation to the host, *J. Exp. Biol.*, **11**(2), 120–139 (1934).

Yetter, W. P.: Codling moth work in Mesa County, *16th Ann. Rep. Sta. Entomol. Colorado, 1924, Cir.* 47, 32–40 (1925).

Chapter 2
Nature of Chemical Attractants

Odors and Tastes. There is one outstanding feature of the mechanism of stimulation by chemicals which characterizes it and sets it apart from that of physical stimulation. Chemical stimulation is accomplished by relatively small material particles which, having emanated from a source, impinge upon the receptors of the animal being affected. From a given source, however distant, molecules are constantly diffusing into the surrounding medium. Under ideal conditions a concentration gradient is established. Whether the stimulating molecules are traveling in a gas or a liquid this gradient exists until all have become evenly distributed throughout the medium and equilibrium attained. Under the environmental conditions in which terrestrial insects exist, equilibrium is attained more frequently in liquids than in the air; consequently, terrestrial insects normally are subjected to a gas (air) in which there are concentration gradients, and to liquids (sap, nectar, fermented juices, etc.) which have attained equilibrium. We are accustomed to referring to stimulating molecules in air as odors, to those in liquids as tastes. Various authors have set forth numerous criteria for distinguishing between taste and smell (Minnich, Parker, Fraenkel and Gunn, von Frisch, von Buddenbrock). Only the more pertinent of these criteria will be pointed out for consideration here. Chemostimuli emanating from distant sources are said to be smelled while those originating from sources in close contact with an animal (within the buccal cavity or touching the legs, palpi, or antennae) are said to be tasted. So-called taste substances usually stimulate one set of receptors while odors affect a different set. Olfactory receptors possess a low threshold and accordingly are extremely sensitive while gustatory receptors are characterized by comparatively high thresholds (cf. Chapter 6).

In the discussions which follow the conception of taste and smell as distinct even though closely related senses is important to remember. The elements of distance and sensitivity especially invite attention.

So little is understood concerning the so-called common chemical sense of insects that at the moment it cannot be intelligently correlated with this discussion. We know simply that certain compounds

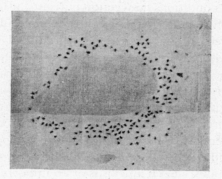

Fig. 6. Fruit flies (*Dacus*) arranged in a concentration gradient of oil of citronella. (From Howlett, *Trans. Entomol. Soc. London*, 1912.)

(e.g., ammonia, chlorine) at high concentrations may be perceived by unidentified receptors located on the legs, cerci, or antennae.

Gradient Fields. An insect placed in a gradient field can, by a series of avoiding reactions at high and low concentrations, finally place itself in a zone of preferred concentration much as it would in a zone of preferred temperature. Ticks will collect in a preferred zone of concentration of butyric acid (Fig. 7); cheese mites (*Tyroly-*

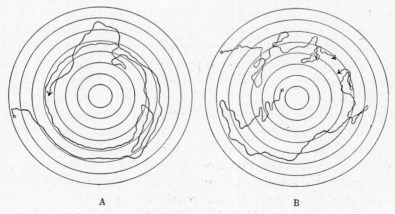

A   B

Fig. 7. Ticks in a concentration gradient of butyric acid. (A) *Ixodes* nymph. (B) *Ixodes* larvae. (Redrawn after Totze, *Z. vergleich. Physiol.*, 1933.)

chus casei) will aggregate at a preferred concentration of skatole, one of the constituents of putrefying protein. Under these conditions directive responses ordinarily are not possible.

Under actual conditions convection currents, usually of air, constantly upset diffusion gradients. Streams of molecules are dispersed from their source. While an insect is stimulated initially by the molecules, subsequent directed reactions are assisted by orientation to a current. The interdependence of odors and currents in eliciting directive responses is illustrated by experiments with *Drosophila*. This fly readily follows up a stream of air bearing the odor of fruit but remains entirely unresponsive to a nonodorous stream. Without convection currents, either extraneous or self-induced, it is unlikely that insects could orient to odors at all. The relation between anemotaxis and responses to chemicals is of an importance that cannot be overemphasized.

Since, by definition, attractants are stimuli which elicit positive directive locomotor responses, true chemical attractants may exist only in the form of odors. Taste stimuli do not permit of positive directive locomotor responses. On the other hand both odors and tastes may serve as repellents since both may call forth simple avoiding reactions.

Humidity Reactions. As Fraenkel and Gunn have pointed out, there is little reason for dissociating hygrostimuli from other chemical stimuli. Reactions to humidity are responses to molecules of a chemical compound, water. The omnipresence of water in all environments is hardly sufficient reason for considering it as a chemical apart. The tendency to erect a special category for humidity reactions on the ground that water affects receptors by physical means is hardly justified since there is a possibility that many other chemicals operate by such physical means as changes in surface tension.

High humidity is powerfully attractive to many arthropods. Anyone who has ever turned over a rotten log or a rock beneath which the earth remains moist must have noticed the large numbers of woodlice (*Porcellio*) gathered there. Gunn (1937) demonstrated experimentally that woodlice aggregate in moist air by an orientation reaction in which there is a rough proportion between the intensity of the stimulus and the frequency of response (orthokinesis). African migratory locusts, on the other hand, seek out dry microclimates

(Kennedy, 1937). Still more striking examples of the attraction of high humidity for some insects and its repellence for others are found among the ground crickets of the genus *Nemobius* of which *N. griseus* seeks very dry habitats while *N. palustris* aggregates where the humidity is highest—in wet sphagnum bogs. Under laboratory conditions houseflies exhibit a marked preference for moderately humid air as against dry or excessively humid air. Excessive moisture also repels *Drosophila melanogaster* while humid air attracts *Lucilia*.

The sensory mechanism associated with reactions to humidity is not at all well understood. Recent work by Lees (1943), in which it was shown that wireworms of the genus *Agriotes* avoid dry air, revealed that the intensity of avoidance is greatest when alternatives are close to saturation. At this point a difference of 7.5 per cent RH is sufficient to cause a response. The beetle *Ptinus tectus*, which reacts to a humidity gradient by collecting in a drier area, exhibits most intense reactions when differences are at low humidities (Bentley, 1944). With regard to *Agriotes*, the intensity of reaction is in better accord with humidity differences when these are expressed as saturation deficiencies rather than as RH. To Lees this suggested that the reaction is initiated by evaporation of water (evaporimeter receptor) and does not involve hygrometer receptors. The site of evaporation appears to be at head appendages, but sensilla conceivably possessing a hygroscopic function are absent. With *Ptinus* the receptors also appear to be located in the head region, especially on the antennae. An increase in the intensity of reaction follows removal of one to five segments. Removal of more than five causes a decline in the intensity of reaction. When only one or more of the proximal three segments remain, the animals show greatly increased activity. As with *Agriotes*, however, no appropriate receptors have been located.

Source of Odors and Tastes in Nature. Aside from water, the most common source of diffusing molecules in nature is organic material, plant and animal. Insects thus react chemopositively and chemonegatively to water, plants, animals, and the products thereof.

In the final analysis all organic matter stems from the green plant. From the raw materials carbon dioxide, water, and inorganic salts, it is able, through the action of sunlight on chlorophyll, to synthesize simple primary compounds, chiefly carbohydrates and amino acids. From these, a vast array of complex secondary prod-

Table 1

SYNTHESIS OF ORGANIC COMPOUNDS IN GREEN PLANTS*

Raw Materials	Primary Products	The More Important Secondary Substances	
		Mol. Wt. less than 1000	Materials of High Mol. Wt.
CO_2 H_2O Inorganic nitrogen compounds	Sunlight on chloroplast pigments {Sugars, Amino acids} Reserve materials Proteins Fats Oils Polysaccharides (e.g., starch) Hemicelluloses	Polyene pigments, aliphatic alcohols, acids, terpenes, sterols, waxes, phosphatides (e.g., lecithin), inositol, aromatic hydroxy acids, hydroxy compounds as glucosides (phenols, complex alcohols), volatile esters, ethers, aldehydes, alcohols (in essential oils with terpenes), alkaloids, pyrrol pigments, anthocyanin pigments, nucleic acids	Cellulose Hemicelluloses Gums Pectins Resins Rubber Tannins Lignins

* From Conant: "The Chemistry of Organic Compounds." By permission of The Macmillan Company, publishers.

Table 2

THE FOOD CYCLE *

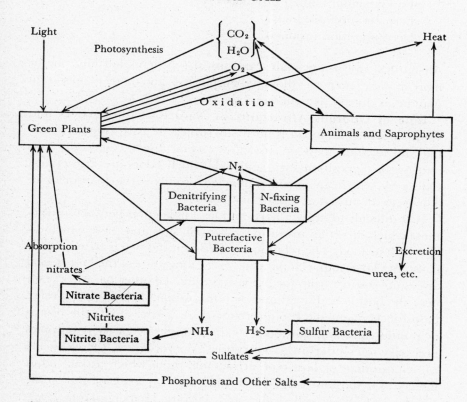

* Reprinted from "Outlines of Modern Biology," by C. R. Plunkett. Copyright, 1929, by Henry Holt and Company, Inc.

ucts is elaborated. The more important categories are listed in Table 1. From this initial supply of synthetic raw materials animals, saprophytes, and various bacteria elaborate still more organic compounds. In turn, all are broken down eventually until the fundamental building blocks are returned to the green plant for further photosynthesis. The cycle of chemicals in nature is known variously as the food cycle, carbon cycle, and nitrogen cycle (Table 2).

Not all of these chemicals, however, are capable of stimulating insects. Surveying the aggregate as a whole one sees that compounds capable of serving as adequate stimuli may be grouped conveniently into four classes:

Fundamental plant products
Fundamental animal products
Fermentation products
Decomposition products

All natural insect-attractants and insect-repellents fall into one or more of these categories though their purpose and mode of operation may be as diverse as their number.

Rôle of Odorous Attractants in Nature. Natural odors and scents to which an insect is normally subjected assist it in recognizing and locating its mate, its fellow (social insects, especially ants and bees), oviposition sites, and food. Such odors are attractants. Odors by which an insect is repelled usually serve to protect it. These are discussed at length in Chapter 8.

It is difficult to draw a hard and fast distinction between attractants which direct an insect to oviposition sites and those which direct it to its food. The relationship between the two is by no means constant. To certain ectoparasites (e.g., mosquitoes) the host is food and quite distinct from the oviposition site. To other parasites (e.g., lice) the host is both food and oviposition site. To still others (e.g., parasitic Hymenoptera, often referred to as parasitoid) the host is oviposition site and not food. In the interest of clarity it becomes necessary to erect arbitrary categories. We recognize, therefore, (1) as ovipositional or host-type attractants those which attract adult insects to oviposition sites which may serve as food for the larvae (insects in case being flesh flies and parasitic Hymenoptera), (2) as food-type attractants, those which attract an insect directly to its food.

It should be borne in mind that substances which are ovipositional-type attractants for one insect may be food-type attractants for another, or for a different stage of the same insect. Such are attrahent principles of plants which direct adults to oviposition sites and larvae to food. Another striking example is to be found in the way in which the products of protein putrefaction are food-type attractants for many necrophagous and coprophagous insects and ovipositional-type attractants for certain cyclorrhaphous Diptera such as blowflies. In general it may be stated that the two great classes of attractants are (1) those which relate to sex and (2) those which relate to food. There are in addition a few natural attractants of doubtful significance and a few powerful man-made attractants which, as far as is known, have no counterpart in nature.

While most attractants are species-specific a few are sex specific also. Naturally, ovipositional-type attractants are effective only with females. Though feeding-type attractants are usually reduced to species specificity only, there are reasons why sex specificity also must exist in certain ones. The food preferences of both sexes are not always identical (e.g., mosquitoes). Sex specificity of feeding-type attractants, however, cannot always be explained on this basis. A few examples, not yet clearly understood, follow: female codling moths are more responsive than males to anethole added to standard bait (Worthley and Nicholas, 1937); *Drosophila* females show a maximum response to ethyl alcohol and acetic acid at concentrations of 10 to 15 per cent and 0.4 per cent respectively while the maximum responses of males occur at 5 and 0.2 per cent respectively (Reed, 1938) (this is clearly a difference in response to odor, not a difference in activity); Wieting and Hoskins (1939) found that female houseflies are more strongly attracted than males to ammonia and vice versa with respect to ethyl alcohol. Sexual differences of response to odors are especially notable in fruit flies. Males of *Ceratitis capitata* are more responsive to kerosene, benzene, and naphtha than are females (Severin and Severin, 1914). Males of *Dacus diversus* and *D. zonatus* are much more attracted to citronella than are females (Howlett, 1912).

Finally, we recognize the fact that no one attractant alone performs the service of guiding an organism to its proper habitat, or mate, or food. The desired end is achieved by a complex array of stimuli working in harmony. In chemical attractants, however, one finds precision guidance. The action of others, as light, temperature, humidity, is comparatively gross, approximating as it does usually, a vicinity. The extreme value of chemical attractants to an organism lies in the specificity and accuracy of operation.

In this regard attractive odors act in the nature of token stimuli. Odors are merely signposts. The beetle *Creophilus* is attracted by odors of decaying meat yet subsists substantially upon fly maggots developing there. As will be seen later, most caterpillars are directed to their food plant by essential oils contained therein yet derive no digestive benefit from these particular constituents of the plant. Bees, also allured by essential oils, feed not on the oils but on nectar.

FIG. 8. Attraction of male gypsy moths to material isolated from female abdomens. The paper cone contains a cotton wad impregnated with the sex attractant. (Courtesy, U.S. Department of Agriculture, Bureau of Entomology and Plant Quarantine.)

Here food odors are truly token stimuli. As von Frisch (1943) has proved, a bee which has located food carries food odors back to the hive on its body. In the dance that follows, other bees are stimulated by this particular odor and depart to search for it. Steinhoffe (unpublished—quoted by von Frisch) found that bee bodies carried odor much better than such materials as glass, porcelain, etc. She demonstrated that a bee conditioned to citral will fly to an isolated abdomen which has lain in citral for one hour and thoroughly examines this abdomen with its feelers.

The token nature of attractants is further attested by the revelation that the attractant alone is not always capable of stimulating feeding or oviposition. Many odors that attract blowflies to oviposition sites will not incite oviposition. Certain odors which guide parasitic Hymenoptera to hosts may not elicit oviposition. Solutions that cause wireworms to aggregate in the vicinity of food fail to initiate biting reactions. In each case another stimulus is required.

Sex Attractants. Early naturalists were amazed and long mystified by the unerring instinct that guided males of certain moths to the waiting virgin female. The phenomenon, known as assembling, was spectacular and mysterious. Distances which males presumably traveled and the fact that they traveled with or at right angles to the wind as well as against it prompted ideas of emanations, waves, and other intangible mechanisms in the minds of those who observed this behavior. Today we realize that sex odors, windborne, enable males to orient from a distance to females of the same species. Orientation of this sort has been described for a number of insects, notably the silkworm moth, *Bombyx mori* (Kellog, 1907); *Arctias selene* (Mell, 1922), the adult mealworm; *Tenebrio molitor* (Valentine, 1931), the gypsy moth; *Liparis dispar* (Collins and Potts, 1932; Prüffer, 1937); species of *Ephestia, Plodia, Galleria,* and *Achroea* (Dickins, 1936; Barth, 1937); two species of European vine moths, *Polychrosis botrana* and *Clysia ambiguella* (Götz, 1939 and 1939a); *Habrobracon juglandis* (Murr, 1930); and *Megarhyssa lunator* (Abbott, 1936).

Scent organs in which odorous sex attractants are elaborated attain their highest development in Lepidoptera, especially Lasiocampidae, Bombycidae, and Saturniidae, in which fully developed eggs are ready for fertilization immediately upon the female's emergence from the pupa (Eidmann, 1929 and 1931). Such glands are

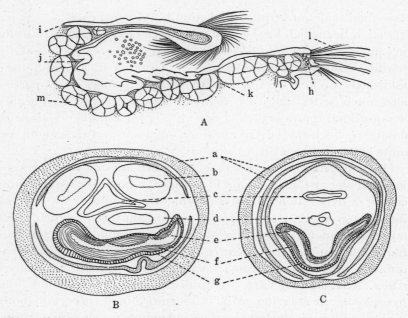

FIG. 9. (A) Diagram of a horizontal section through the last two segments of a female *Euploea asela:* (h) gland cells, (i) cuticle, (j) hypodermis, (k) blood, (l) glandular hairs, (m) fat body. (Redrawn after Freiling, *Z. wiss. Zool.*, 1909.) (B and C) Diagrams of cross sections of the ovipositors of *Ephesta cautella* and *E. kühniella*, respectively, showing the location of odor glands: (a) body wall of segment seven, (b) odor gland, (c) rectum, (d) oviduct, (e) intersegmental fold containing gland cells, (f) segment eight, (g) glandular epithelium. (Modified after Dickins, *Trans. Entomol. Soc. London*, 1936.)

located adjacent to the external genitalia where they are visible as tufts of modified scales, hairs, or evaginations of the body wall in which the glandular epithelium is covered by a thin cuticle permeable to the secretion (Fig. 9) (Freiling, 1909). Only when the virgin female assumes a so-called "calling pose" is the attractive scent dispersed. Typically the "calling pose" is signaled by eversion of the scent organ directly or as a result of extension of the genitalia. Immediately upon locating a female the male orients for copulation by contact whereupon the female scent organs may be withdrawn. That this glandular secretion is the source of the attracting odor is adequately attested by the fact that males will attempt to copulate with pieces of blotting paper and other materials which have been wetted with secretion. Also, males of the hymenopteron *Habrobracon* may be attracted to other males that have but recently copulated, indicating that even such minute amounts of secretion as may contaminate males in coitu are attractive.

Sex scents are dispersed by wind, and the distances over which they are effective depend upon wind velocity and direction, relative humidity, temperature, and other meteorological factors. In the case of June beetles (*Phyllophaga lanceolata* (Say)) odors are especially effective in sunshine and moderately strong breezes. Numerous ingenious experiments have been conducted to ascertain the maximum distances over which these odors are operative. Marked males of *Arctias selene* were discovered to be orientating from a distance of 11 kilometers (Mell, 1922). Traps baited with female gypsy moths and located on treeless islands caught males assembling from as far away as 2$\frac{5}{16}$ miles though average flights ranged from $\frac{1}{4}$ to $\frac{1}{2}$ mile (Collins and Potts, 1932). Travis (1939) found that in the absence of wind the sphere of attraction of female June beetle scent may be as great as 15 to 20 feet; with a moderate breeze, 30 to 40 feet on the windward side; with a strong wind, 50 to 75 feet. In the latter instance, however, few males succeed in reaching the source of the odor against the force of the wind.

Extensive experiments with European vine moths indicated, as was to be expected, that the degree of attraction increases with concentration. It varies also with the age of the insects. The relative attractiveness of this scent was tested by baiting traps with live virgin females and erecting, for comparison, traps baited with the best artificial attractants known at the time, a wine made from marc of grapes plus 2 per cent vinegar and 1 per cent sugar. For *Polychrosis botrana* and *Clysia ambiguella* respectively, 38 and 52 times as many males were caught in the female-baited traps as in the controls. Such a comparison, of course, is not especially significant as a test of the relative attractiveness of this specific scent since two basically different types of attractants are being compared, sex and food-type attractants. Here is merely indication that under special conditions sex responses take precedence over feeding responses.

Theoretically, female scent is species-specific. In the field this usually holds true; in confinement, however, it may not be so. Dickins (1936) and Barth (1937) reported that males of *Ephestia*, *Plodia*, *Galleria*, and *Achroea* are attracted and attempt to copulate interspecifically when caged.

Chemistry of Sex Scents. The chemistry of sex scents is virtually unknown. Not the least of difficulties hindering successful analyses is the minute quantity of material available for examination. Probably the most thorough and exhaustive analyses yet attempted were those of Collins, *et al.*, and Haller and his colleagues. Nevertheless, the results of this extensive work were largely negative in that the substance remains unidentified. Collins, *et al.*, concluded that the scent substance from female gypsy moths is probably a lipoid, protein or ester and is saturated. It is soluble in alcohol as well as in ordinary fat solvents; it appears in unsaponifiable fractions and is neither an acid, base, nor aldehyde. It is specific for the male gypsy moth. Haller, *et al.*, are in agreement except on the point of saturation. They found that the attractiveness of the substance is markedly increased by treatment with catalytic hydrogen. They further determined that the attractant is not cholesterol.

A less direct attempt at solving the problem of the nature of sex scents is a hit or miss testing of suspected compounds. Such an approach was suggested by the early work of Howlett whereby he discovered that oil of citronella is very attractive to males of the fruit flies *Dacus diversus* and *D. zonatus*. He offered a suggestion that oil of citronella approximated in action the odorous secretion of females. Later he found that males of *D. diversus*, *D. zonatus*, and *D. ferrugineus* are attracted respectively to isoeugenol, methyleugenol, and to both iso- and methyleugonol, but no further proof was forthcoming to indicate whether these chemicals simulated food odors or sex odors. The fact that males alone appeared to be attracted to mango, guava, etc., and that these plants possessed odors similar to that of eugenol derivatives, suggested that the eugenols might be food-type attractants for males only. On the other hand, Ripley and Hepburn (1931) accumulated considerable evidence relative to the responses of the Natal fruit fly to male attractants to support their "female-odor" hypothesis. Briefly, these are the salient facts: (1) both sexes are strongly attracted by odors emanating from fermenting carbohydrates, the principal food; (2) only males are attracted by terpinyl acetate, caryophyllene, kerosene, diphenyl methane, isopulegyl formate, and dihydrocarveol acetate and then only under field conditions; (3) females are repelled by these chemicals in the field and both sexes are strongly repelled by them under laboratory conditions; (4) these chemicals are not known to occur in the animal kingdom and the last three enumerated are not known to occur in nature. The only theory consistent with these primary facts and all supplementary facts is that the male sensory system confuses these odors with female scent. In short, the female-odor hypothesis postulates that these odors imitate female scents.

Tests with approximately 150 aromatic compounds (Lehman, 1932) indicated that the fatty acids, especially butyric, caproic, valeric, and lactic may be sex attractants for the adult wireworms *Limonius canus* Lec. and *L. californicus* Mann. since in the field they attracted a majority of males and in the laboratory caused males to extend their genitalia. Ethyl pelargonate likewise was very attractive. Collins and Potts, in their tests with gypsy moths, found that allylamine, ethylamine, and methylamine are attractive to males. The degree of attractiveness decreased in the order named. Travis (1939), seeking potential attractants for the June beetle, tested methyl salicylate, ethyl butyrate, methyl-phenyl acetate; saligenin, isoamyl mercaptan, vanillyl alcohol, allyl alcohol, piperonal, isoamyl alcohol, heptaldehyde, benzaldehyde, anisaldehyde, anisol, ethyl benzoate, ethyl anisate, eugenol, isoeugenol, eugenol methyl ether, oil of wintergreen, and isoamylamine. The last named, isoamylamine, elicited the same response from males as did crushed or sexually active females.

Although the evidence is much too scanty to form a basis for any generalizations, it appears that in some species sex odors may originate with fatty acids and amines. Just how specific the amines enumerated are in their action remains to be learned. Here is a field concerning which we must still profess a great ignorance.

Recognition Odors. Still greater is our ignorance concerning odorous compounds by which insects recognize their nests and nestmates and follow trails. That such odors exist is unquestionable. Gregarious larvae of *Pieris brassicae* detect species odor from a distance of 5 cm. Ants that have been smeared with the body juices of intruders or have simply been washed are themselves promptly treated as intruders by their nestmates. Social insects live in a world not only of individual odors but also of caste odors, hive or nest odor, sting odor and, in the case of bees, wax odor. By training himself, McIndoo was able to detect caste odors and distinguish members of different castes by this character. His review paper (1917) treats of the subject exhaustively and contains a complete bibliography. No conclusive chemical studies have been undertaken.

Though probably closely related chemically to sex attractants, recognition scents and aphrodisiac scents are probably not true attractants.

Ovipositional-type Attractants. Inasmuch as the chemical factors which operate in the selection of oviposition sites are frequently identical with those governing feeding relations and operate in much the same manner, it is unnecessary at this point to do more than select several of the more interesting examples to illustrate this type of reaction. Specific examples will be discussed at greater length together with the feeding-type attractants with which they are identical.

As Richardson (1925) pointed out, the external stimuli governing the time and location of oviposition include temperature, humidity, light, currents, surfaces, odorous substances, and contact with chemical substances. From the point of view of attractants, odors are the most important of these. Not only do they attract gravid females, but many, operating alone, actually induce oviposition. Those groups of insects concerning which most experimental work has been attempted include: parasitic Hymenoptera, entomophagous Diptera, entomiasis-producers, bloodsucking species, saprophagous Diptera, and phytophagous insects.

Parasitic Hymenoptera are outstanding for the apparent ease with which they are able to locate insect hosts upon which to oviposit. One need but recall the accuracy with which the ichneumonid *Megarhyssa atrata* bores into trees to oviposit in woodboring larvae of *Tremex columba*, the ability of *Rhyssa* to recognize larvae of *Sirex* through several centimeters of wood, and the ability of *Trichogramma* to detect those host eggs that have already been parasitized to appreciate the existence of some guiding principle. Host specificity among Hymenoptera of this group is the rule rather than the exception. The chalcid *Embidobia* develops in the eggs of Embioptera; the proctotrypid *Polygnotus*, in the larvae of Cecidomyidae; species of *Scelio*, in locust eggs only, the braconid *Dacnusa* in larvae of Agromyzidae (Diptera); *Aphidius* and allied braconids in Aphididae; the braconid *Perilitus* in adult Coleoptera (Imms, 1931). Up to a certain point chance alone determines whether or not a host is to be selected. Thence the attrahent principle is host odor. It is subtle and specific. To within a certain distance from the source the insect moves by apparently random or klinokinetic methods. As the odor becomes stronger the insect makes a final dash in a more or less straight line. Thus *Habrobracon* when approximately 2 cm. from its host, *Ephestia*, advances directly to the caterpillar. This is probably klinotaxis. It has been shown that the following also locate and select their hosts for oviposition by odor: *Microplectron fuscipenne* Zett. (Chalcididae) (Ullyett, 1936), parasites of the meal-moth larva, *Ephestia* (Thorpe and Jones, 1937), and *Trichogramma evanescens* (Laing, 1938). *Trichogramma*, an egg parasite, is even able to distinguish untouched eggs from those already parasitized. The precise experiments of Thorpe and Jones, in which an olfactometer was employed to prove the rôle of odor in attraction, showed that in the case of *Nemeritis canescens* (Grav.) (Hymenoptera, Ichneumonidae) host odor and to a small extent the odor of the host's food (meal) is the attractant. In Europe this insect is parasitic exclusively on larvae of species of *Ephestia*. In the new world a race of *Nemeritis* attacks the large wax moth, *Galleria mellonella* (Zell.). The European race may be reared with difficulty on larvae of *Galleria* and more easily on larvae of the small wax moth, *Meliphora grisella* (F), provided the latter is placed with *Ephestia* larvae to become saturated with that host odor. Preferences for *Ephestia* odor are germinally fixed, but species reared on the abnormal host *Meliphora* are attracted secondarily to that odor. Some olfactory conditioning to the new host occurs in the pre-imaginal period; nevertheless, the odor of *Ephestia* is still preferred. Neither the source nor chemistry of host odor is definitely known.

Among the flies, entomophagous parasites are far less restricted than Hymenoptera, though exceptions may be cited. The related dexiine genera *Thelaira, Fortisia, Cyrillia,* and *Thrixion* attack lepidopterous larvae, the chilopod *Lithobius*, the isopod *Metaponasthru*, and Phasmidae, respectively (Imms, 1931). The vast group of tachinid flies conform to the rule, however, and are preëminently polyphagous. Host relations in this group may be extremely complex, for in addition to ovipositing directly on the host these parasitoid insects may lay eggs on the host's food in which case they are swallowed with the food, or merely in the vicinity of the host where the aggressive larvae seek out a host. The tachinid *Gonia capitata*, parasitic on a cutworm, *Porosagrotis orthogonia*, exemplifies the former relationship. A clue to stimuli in cases of this sort appears in the observation of Townsend (1908) to the effect that *Eupeleteria magnicornis* Zett. deposits larvae only on that portion of the host's food plant as has already been traversed by the host. This suggests host odor. In the absence of experimental work bearing upon parasitic Diptera we must assume tentatively that in direct host-parasite relationships the same factors operate as with parasitic Hymenoptera.

A host of Diptera, chief among which are the bot flies, warbles, blowflies and flesh flies, seek a host, usually vertebrate, for the purpose of laying eggs or larvae thereon. Largely on account of the great economic importance of many of these insects a considerable amount of work has been directed toward the prevention of oviposition on cattle and man.

As Folsom and Wardle (1934) have pointed out, it is difficult to distinguish between habitual and accidental entomiasis. The former arises from the latter. Probably *Lucilia sericata, Chrysomyia albiceps, C. varipes, Anastellorhina augur, Pollenia stygia* are primarily carrion breeders which attack live animals only secondarily. *Cochliomyia americana* in North and Central America probably represents a transition stage between true carrion breeders and actual parasites. It oviposits in wounds and natural orifices. True entomiasis is represented by *Chrysomyia bezziana* and *Wohlfartia magnifica*. Habitual entomiasis-producers are to be found in the families Muscidae, Oestridae, Sarcophagidae, and Sarcopsyllidae (fleas).

Host odor is obviously the prime attractant. Closely associated is the odor of putrefying protein material. Vertebrate host odors have been more thoroughly analyzed than those of invertebrates and found to be due to the secretions of various glands, to clotting or dried blood, and to putrefying flesh in the vicinity of wounds. In a broad sense the odors are similar to those originating from carrion and dung. Most arise from the multitudinous products resulting from the breakdown of proteins and fats. It goes without saying that the flies enumerated above are attracted to various baits composed of meat, blood, or carrion. Thus *Calliphora* will follow a randomlike course to within 6.4 cm. of meat at which point it flies directly to its destination (Hartung, 1935); blowflies are strongly attracted to a mixture of 50 g. of fresh liver and 20 cc. of water, as well as to soups prepared by allowing trapped blowflies to putrefy in water for five days (Freney, 1932). Various animal tissues *per se* are not especially attractive until they have begun to putrefy. Attempts to trap blowflies with highly attractive compounds extracted from carrion baits have not been attended with marked success. According to Freney (1937), the only chemical attractive in pure state is ethyl mercaptan; however, others such as indole, skatole, and ammonium carbonate do exert some attraction (Hobson, 1936 and 1937); and the addition of sodium sulfide, acetic acid, alkyl sulfides, or calcium carbonate appreciably increases the attractiveness of baits (Fuller 1934; Freney 1937).

Concerning the ovipositional-type attractants of blood-sucking species that pass their developmental stages in water (Simuliidae, Culicidae, Ceratopogoninae (punkies), Tabanidae) little is known. Most of the work pertains to mosquitoes. Sharma and Sen (1921) reported that weak solutions of sodium chloride, sodium citrate, and sodium tartrate are conducive to egg laying. Corresponding acids are repellent. The addition, also, of hydrogen sulfide, methane, or stale urine to water is said to encourage oviposition by *Culex pipiens.*

Not a few of the insects associated with decaying organic material congregate at such centers for the dual purpose of feeding and egg laying. Roubaud and Veillon (1922) and Richardson (1925) state, however, that the house fly is attracted to manure for oviposition only. Here the chief attractants are ammonia or ammonium carbonate plus butyric or valerianic acid.

Adult phytophagous insects generally are attracted for purposes of oviposition by the same compounds serving as feeding-type attractants for immature stages. The corn ear worm moth, for example, will lay eggs on cotton twine soaked with fresh corn silk juice (McColloch, 1922). Here surface characters also are important. Taste likewise may be a deciding factor in some instances. According to Hancock (1904), one of the meadow grasshoppers (*Orchelimum glaberrimum* Burmeister) briefly tastes a plant before laying any eggs on it. Oviposition by citrus borers (e.g. *Citripestis sagittiferella*) is induced by the odor of citral, the principal ingredient of oil of limes (Pagden, 1931). The lesser grain borer (*Rhizopertha dominica*) locates its larval food by odor and, once within the odor gradient field, substitutes the sense of touch for determination of an egg laying site. It may also be conditioned to foreign odors such as peppermint (Crombie, 1941). Lipp (1928 and 1929) found that a 2 per cent alcoholic solution of allyl sulfide attracted adults of the Japanese beetle which had been feeding on plants containing sulfides and induced them to oviposit freely in sod adjacent to a dish containing the mixture.

Food-type Attractants. The most important and elemental attractants by far are those which guide an insect to its food. In every case, in the final analysis, odor is the organism's index regardless of the type of food.

In discussing feeding habits it is customary to group insects on the basis of the kinds of food eaten. Four basic categories may be recognized, each with its various subdivisions, but the number and scope of subdivisions receive different treatment from different authors. The scheme presented below is a composite.

1. Zoophagous (carnivorous)
    Entomophagous (parasitic)
    Harpactophagous (predatory)
2. Saprophagous (scavenger)
    Coprophagous (feeding on excreta)
    Necrophagous or Saprozoic (carrion feeding)
    Saprophagous (feeding on disintegrating plant tissue)
    Geophagous (feeding on earth containing organic matter)
    Detrivorous (feeding on fur, feathers, etc.)
3. Pantophagous (omnivorous)
4. Phytophagous (vegetarian or herbivorous)

So large is the group of plant-feeding insects and so complex the interrelations of insects and plants that it becomes necessary further to subdivide the group before any concise discussion be attempted.

Plant-feeders may be classified by using as criteria either the phylum or the anatomical part of the plant upon which they subsist. As the majority of cryptogams do not exhibit highly specialized anatomical features comparable to the leaves, inflorescence, and roots of higher plants and as those insects feeding on cryptogams usually attack the whole plant indiscriminately, a taxonomic classification ordinarily suffices. On the other hand, insects attacking phanerogams characteristically select a limited portion of the plant for feeding purposes. By combining the two systems of classification judiciously, we may arrive at a working outline which encompasses the more important categories to be considered in this chapter. Lichen-feeders (lichenophagous), moss-feeders, and fern-feeders are purposely omitted here and in subsequent discussions because nothing is known of the attractants or repellents influencing the behavior of these insects.

Phytophagous insects
   Alga-feeders (phycophagous)
   Fungus-feeders (mycetophagous)
   Spermatophyte-feeders
     Leaf-feeders
      Surface-feeders
      Miners
     Fruit-feeders (carpophagous)
      Those feeding on fresh fruit
      Those feeding on decaying fruit
     Flower-feeders
     Wood-feeders (xylophagous)
      Stalk-feeders
      Root-feeders
      Wood-borers

When attempts were made to present the story of feeding-type attractants by following the aforementioned outline in the usual manner, it immediately became apparent that such a scheme, if strictly adhered to, was impractical and artificial. There is too much overlapping and too many insects with nutritional habits of a doubtful nature.

Nowhere is this better illustrated than in the associations of mycetophagous insects and fungi, associations which are, for the most part, little understood. Lacunae in our knowledge are due in no small measure to the wide dissimilarity of habitats of fungi, to the multiplicity of forms and sizes, to the universal presence of fungi on all kinds of substrata. Feeding relations between insects and large fungi usually permit of direct interpretation. It is the unobtrusive presence of microscopic forms everywhere that has for a long time obscured the true picture of the feeding habits of many arthropods. It now appears that numerous insects actually subsist upon fungi rather than upon the various substrata which were formerly believed to be their food. Many coprophagous insects feed upon coprophytes for which feces serves merely as a growth medium; numbers of xylophagous insects feed on wood-growing fungi; a large number of insects subsisting on spoiled, soured, or decaying fruit are in reality dependent upon associated fungi.

It is not always easy to determine whether an insect feeds upon a microscopic fungus by consuming the substratum in bulk, whether ingestion of the fungus is incidental to ingestion of the substratum, whether the fungus alone is ingested, or whether products of the action of the fungi on the substratum constitute the food. For example, many wood-boring insects are xylophagous in the true sense of the word, others consume decaying wood for the fungi growing thereon, while still others, such as certain Ambrosia beetles, subsist entirely on associated fungi. Similar situations are to be found among insects feeding on feces, dead or decaying organic material, and spoiling fruit. In view of these facts, it has become customary to describe such insects as microphagous. From the point of view of attractants it is more desirable to group them as in the following outline. The chief justification for this system lies in the fact that attractants associated with microflora are usually products of the action of divers fungi on the substratum and as such are frequently more characteristic of the substratum than of the species or even genus of fungi (cf. p. 78). Thus different, attractive, intermediate, breakdown products are characteristic of each kind of organic material, though the agent responsible in each case could conceivably be the same.

As has been repeatedly pointed out, attractants are token in nature. It is clearly more logical to group the species under discussion fundamentally on the basis of attractants. Experience indicated that chemical food-type attractants fall more or less naturally into three well-defined groups whose natures will be examined at length in subsequent chapters. For the present the following outline sets forth the subject as it will be presented.

Fundamental plant products (essential oils, resins, and related substances)
  Leaf-feeders
  Flower-feeders
  Certain fruit-feeders
  Some xylophaga
  Some sap-feeders
Fermentation products (alcohols, acids, aldehydes, esters, and carbinols)
  Carpophaga
  Xylophaga

Sap-feeders
Mycetophaga
Fundamental animal products and protein and fat decomposition
    products (fatty acids, amines, ammonia, carbon dioxide)
Entomophaga
Harpactophaga
Coprophaga
Necrophaga
Saprophaga
Geophaga
Detrivera
Phycophaga
Mycetophaga

The value of this system will become increasingly apparent as we progress. Suffice it to mention at this point that the grouping of such insects as mycetophagous and necrophagous or harpactophagous and phycophagous is not so alien as it may seem. Examples of such insects being attracted to foods of widely divergent origin are by no means rare. Likewise, it is apparent that a given compound may appear in more than one of these categories. Oxalic acid, for example, is a so-called fundamental product of many plants, especially the Polygonaceae, yet it is to be found also as a fermentation product following the action of many fungi, e.g., some species of *Penicillium* and *Aspergillus*, and a few in the genera *Polyporus*, *Corticium*, etc. Many alcohols, aldehydes, etc., synthesized by a green plant and classed as components of an essential oil may appear elsewhere in nature as products of fermentation. The attempt has been made in each case to consider the mode of origin of the compound in discussing it in the appropriate chapter. It is to be hoped that further work will clarify the issue by making available more facts through which a sound system of classification may be erected.

Attractants of Doubtful Significance. Smoke. Ordinarily smoke acts as an effective repellent, but in the lives of some buprestid beetles it apparently subserves a useful purpose as an attractant. As early as 1885 Ricksecker recorded an aggregation of hundreds of *Melanophila consputa* Lec. and *M. longipes* Say to an area of burned Douglas firs. Since then numerous instances of this sort have been observed. The most logical behavior in this realm of strange association has been responses to burning conifers or substances of coniferous origin. Swarming by *M. acuminata* (DeG.), *M. notata* (Cast. and Gory), and *M.* sp. to burning pine stumps has been reported (Manee, 1913; Sharp, 1918) as has also swarming by *M. ignicola* Champ. to a distillation plant which produced tar from pine stumps (Champion, 1918). Of more anomalous nature have been aggregations of buprestids to a 750,000-barrel oil fire in a region more than 50 miles from the nearest conifers (Van Dyke, 1926), to a smelter plant (Linsley, 1933), to a sugar refinery (Van Dyke, 1928), and to football games. Linsley (1943), who reported the last case, pointed out that the smoke from "some twenty thousand (more or less) cigarettes" hangs like a haze over the stadium during "big" games.

Linsley has reviewed and carefully investigated the phenomenon of attraction to fire. Of the three possible attractants present, smoke (and associated fumes—usually acrid—including innumerable distillation products), heat, and light, smoke is undoubtedly the prime mover. It acts as an olfactory stimulus, but just what component or components are directly responsible for stimulation remains to be discovered. Heat merely stimulates activity at the scene, and light is unimportant.

As to the possible selective value of this odd habit in the survival of the species, Linsley suggested that it directs beetles to the scene of forest fires where they normally mate and oviposit in scorched coniferous wood. The fire provides a convenient source of freshly killed wood, and the resultant ashes and charcoal may serve as a repellent to competing species.

MAN-MADE ATTRACTANTS. From time to time observations are recorded of the attraction of insects to synthetic compounds which apparently bear no relation to the ecology of the species concerned. More likely than not, our own incomplete understanding of the ecology contributes to our inability to appreciate the rôle of the attractant involved. In nature some of the aromatic unsaturated homocyclic compounds, especially the coal tar crudes, benzene (C_6H_6), toluene ($C_6H_5CH_3$), xylene ($C_6H_4(CH_3)_2$), naphthalene ($C_{10}H_8$), and anthracene ($C_{14}H_{10}$), are very rare. Why the fruit fly *Ceratitis capitata*, especially the male sex, should be attracted to benzene, naphtha, and kerosene is certainly a mystery (Severin and Severin, 1914 and 1914a) as is also the reason behind the attractiveness of high boiling fractions of petroleum oils to females of the chironomid *Forcipomyia* sp. (Ahmad, 1934). Ordinarily compounds of this type are strongly repellent. Other inexplicable cases include the attraction of the palmetto weevil, *Rhynchophorus cruentatus* Fab., to fumes of banana oil (Bare, 1929), that of the predatory wheel bug, *Arilus cristatus* L., to turpentine (Metzger, 1928), and the attraction of kelp flies to the cleaning fluid trichlorethylene (Williams, 1943). Here, as in the other examples given, there is no apparent correlation between the attractant and the habitat of the insect. Kelp flies frequent and breed in drying and decaying kelp, especially *Fucus*, cast up on the beach. Trichlorethylene has not been isolated from this material.

REFERENCES

Abbott, C. E.: On the olfactory powers of *Megarhyssa lunator* (Hymenoptera: Ichneumonidae), *Entomol. News*, **47**(10), 263–264 (1936).

——: The physiology of insect senses, *Entomologica Am.*, **16**(4), 225–280 (1937).

Ahmad, T.: Chemotropic response of a chironomid fly (*Forcipomyia* sp.) to petroleum oils, *Nature*, **133**(3360), 462–463 (1934).

Bare, C. O.: *Rhynchophorus cruentatus* Fab., the Palmetto weevil, attracted to automobile paint, *J. Econ. Entomol.*, **22**(6), 986 (1929).

Barth, R.: Herkunft, Wirkung und Eigenschaften des weiblichen Sexualduftstoffes einiger Pyraliden, *Zool. Jahrb. Zool. Physiol.*, **58**(3), 297–329 (1937).

——: Bau und Funktion der Flügeldrüsen einiger Mikrolepidopteren, *Z. wiss. Zool.*, **150**(1), 1–37 (1937a).

Bentley, E. W.: The biology and behavior of *Ptinus tectus* Boie (Coleoptera, Ptinidae), a pest of stored products, *J. Exp. Biol.*, **20**(2), 152–158 (1944).

von Buddenbrock, W.: "Physiologie der Sinnesorgane und des Nervensystems. Grundriss der vergleichenden Physiologie, I.," 2d ed., Berlin, Gebrüder Borntraeger, 1937.

Champion, H. G.: A note on the habits of a *Melanophila* (Buprestidae) and other Indian Coleoptera, *Ent. Monthly Mag.*, **54**, 199–200 (1918).

Collins, C. W., and S. F. Potts: Attractants for the flying gypsy moths as an aid in locating new infestations, *U.S. Dep. Agr. Tech. Bull.*, 336, 1–43 (1932).

Conant, J. B.: "The Chemistry of Organic Compounds," New York, The Macmillan Company, 1938.

Crombie, A. C.: On oviposition, olfactory conditioning and host selection in *Rhizopertha dominica* Fab. (Insecta, Coleoptera), *J. Exp. Biol.*, **18**(1), 62–79 (1941).

Dickins, G. R.: The scent glands of certain Phycitidae (Lepidoptera), *Trans. Entomol. Soc. London*, **85**(14), 331–362 (1936).

Eidmann, H.: Morphologische und physiologische Untersuchungen am weiblichen Genitalapparat der Lepidopteren. I. Morphologischer Teil, *Z. angew. Entomol.*, **15**(1), 1–66 (1929).

——: Morphologische und physiologische Untersuchungen am weiblichen Genitalapparat der Lepidopteren. II. Physiologischer Teil, *Ibid.*, **18**(1), 57–112 (1931).

Fabre, J. H.: "The Life of the Caterpillar," New York, Dodd, Mead & Co., 1916.

Folsom, J. W., and R. A. Wardle: "Entomology With Special Reference To Its Ecological Aspects," 4th ed., Philadelphia, The Blakiston Company, 1934.

Fraenkel, G. S., and D. L. Gunn: "The Orientation of Animals," Oxford University Press, 1940.

Freiling, H. H.: Duftorgane der weiblichen Schmetterlinge nebst Beiträge zur Kenntniss der Sinnesorgane auf dem Schmetterlingsflügel und der Duftpinsel der Männchen von *Danais* und *Euploca*, *Z. wiss. Zool.*, **92,** 210–290 (1909).

Freney, M. R.: The chemical treatment of baits for attracting blowflies, *J. Council Sci. Ind. Research*, **5,** 94–97 (1932).

——: Studies on the chemotropic behavior of sheep blowflies, *Australia, Council Sci. Ind. Research, Pamphlet* 74, 1–24 (1937).

von Frisch, K.: Vergleichende Physiologie des Geruchs-Geschmacksinnes, *Handb. norm. path. Physiol.*, **2,** 203–239 (1926).

——: Versuche über die Lenkung des Bienenfluges durch Duftstoffe, *Naturwissenschaften*, **31**(39/40), 445–460 (1943).

Fuller, M. E.: Sheep blowfly investigations. Some field tests of baits treated with sodium sulfide, *J. Council Sci. Ind. Research*, **7,** 147–149 (1934).

Götz, B.: Über weitere Versuche zur Bekämpfung der Traubenwickler mit Hilfe des Sexualduftstoffes, *Anz. Schädlingskunde*, **15,** 109–114 (1939).

——: Untersuchungen über die Wirkung des Sexualduftstoffes bei den Traubenwicklern *Clysia ambiguella* und *Polychrosis botrana*, *Z. angew. Entomol.*, **26**(1), 143–164 (1939a).

Gunn, D. L.: The humidity reactions of the woodlouse, *Porcellio scaber* (Latreille), *J. Exp. Biol.*, **14**(2), 178–186 (1937).

Haller, H. L., F. Acree, and S. F. Potts: The nature of the sex attractant of the female gypsy moth, *J. Am. Chem. Soc.*, **66**(10), 1659–1662 (1944).

Hancock, J. L.: The oviposition and carnivorous habits of the green meadow grasshopper (*Orchelimum glaberrimum* Burmeister), *Psyche*, **11,** 69–71 (1904).

Hartung, E.: Untersuchungen über Geruchsorientierung bei *Calliphora erythrocephala*, *Z. vergleich. Physiol.*, **22**(2), 119–144 (1935).

Hartwell, R. A.: A study of the olfactory sense of termites, *Ann. Entomol. Soc. Am.*, **17**(2), 131–162 (1924).

Henschel, J.: Reizphysiologische Untersuchungen an der Käsemilbe *Tyrolichus casei* (Oudemans), *Z. vergleich. Physiol.*, **9**(5), 802–837 (1929).

Hobson, R. P.: Sheep blow-fly investigations. III. Observations on the chemotropism of *Lucilia sericata* Mg., *Ann. Applied Biol.*, **23**(4), 845–851 (1936).

——: Sheep blow-fly investigations. IV. The chemistry of the fleece with reference to the susceptibility of sheep to blow-fly attack, *Ibid.*, **23**(4), 852–861 (1936a).

——: Sheep blow-fly investigations. V. Chemotropic tests carried out in 1936, *Ibid.*, **24**(3), 627–631 (1937).

Howlett, F. M.: The effect of oil of citronella on two species of Dacus, *Trans. Entomol. Soc. London*, **1912,** 412–418.

——: Chemical reactions of fruit-flies, *Bull. Entomol. Research*, **6**(3), 297–305 (1915).

Huffaker, C. B., and R. C. Back: A study of methods of sampling mosquito populations, *J. Econ. Entomology*, **36**(4), 561–569 (1943).

Imms, A. D.: "Recent Advances in Entomology," 2 ed., Philadelphia, The Blakiston Company, 1931.

Kellog, V. L.: Some silkworm moth reflexes, *Biol. Bull.*, **12**(3), 152–154 (1907).

Kennedy, J. S.: The humidity reactions of the African migratory locust *Locusta migratoria migratoroides* R. and F., gregarious phase, *J. Exp. Biol.*, **14**(2), 187–197 (1937).

Laing, J.: Host-finding by insect parasites. II. The chance of *Trichogramma evanescens* finding its hosts, *Ibid.*, **15**(3), 281–302 (1938).

Lees, A. D.: On behavior of wireworms of the genus *Agriotes* Esch. (Coleoptera, Elateridae). I. Reactions to humidity, *Ibid.*, **20**(1), 43–60 (1943).

Lehman, R. S.: Experiments to determine the attractiveness of various aromatic compounds to adults of the wireworms, *Limonius* (*Pheletes*) *canus* Lec. and *Limonius* (*Pheletes*) *californicus* Mann., *J. Econ. Entomology*, **25**(5), 949–958 (1932).

Linsley, E. G.: Some observations on the swarming of *Melanophila*, *Pan-Pacific Entomologist*, **9**(3), 138 (1933).

——: Attraction of *Melanophila* beetles by fire and smoke, *J. Econ. Entomology*, **36**(2), 341–342 (1943).

Lipp, J. W.: A suggestion for a possible application of ovipositional chemotropism, *Ibid.*, **21**(6), 939 (1928).

——: Notes on experiments on ovipositional chemotropism, *Ibid.*, **22**(5), 823–824 (1929).

McColloch, J. W.: The attraction of *Chloridea obsoleta* Fabr. to the corn plant, *Ibid.*, **15**(5), 333–339 (1922).

McIndoo, N. E.: Recognition among insects, *Smithsonian Inst. Pub.*, *Misc. Collections*, **68**(2), 1–78 (1917).

Manee, A. H.: Observations on Buprestidae at Southern Pines, North Carolina (Coleop.), *Entomol. News*, **24**(4), 167–171 (1913).

*Mell, R.: "Biologie und Systematik der südchinesischen Sphingiden," 1922.

Metzger, F. W.: Turpentine oil as an attractant of the wheel bug (*Arilus cristatus* L.), *J. Econ. Entomol.*, **21**(2), 431–432 (1928).

Minnich, D. E.: The chemical senses of insects, *Quart. Rev. Biol.*, **4**(1), 100–112 (1929).

Murr, L.: Über den Geruchssinn der Mehlmottenschlupfwespe Habrobracon juglandis Ashmead, *Z. vergleich. Physiol.*, **11**(2), 210–270 (1930).

Nicholson, G. W.: *Melanophila acuminata* DeG. at a fire in June, *Entomol. Monthly Mag.*, **55**, 156–157 (1919).

*Pagden, H. T.: Two citrus fruit borers, *Dept. Agr. Straits Settlements Federated Malay States, Sci. Ser.* No. 7, 1–16 (1931).

Parker, G. H.: "Smell, Taste, and Allied Senses in the Vertebrates," Philadelphia, J. B. Lippincott Co., 1922.

——, and E. M. Stabler: On certain distinctions between taste and smell, *Am. J. Physiol.*, **32**(4), 230–240 (1913).

Plunkett, C. R.: "Outlines of Modern Biology," New York, Henry Holt & Co., 1930.

Prüffer, J.: Dalsze badania nad zjawiskiem wabienia samców u *Lymantria dispar* L., *Zool. Poloniae*, **2**(1) (1937).

Reed, M. R.: The olfactory responses of *Drosophila melanogaster* Meigen to the products of fermenting banana, *Physiol. Zoöl.*, **11**(3), 317–325 (1938).

Richardson, C. H.: The oviposition responses of insects, *U.S. Dep. Agr. Bull.* 1324, 1–17, (1925).

Ricksecker, L. E.: (in Notes and News), *Entomologica Am.*, **1**, 96–98 (1885).

Ripley, L. B., and G. A. Hepburn: Further studies on the olfactory reactions of the Natal fruit-fly, *Pterandrus rosa* Ksh., *Union S. Africa, Dep. Agr., Mem.* 7, 24–81 (1931).

Roubaud, E., and R. Veillon: Recherches sur l'attraction des mouches communes par les substances de fermentation et de putrefaction, *Ann. inst. Pasteur*, **36**, 752–764 (1922).

Salt, G.: The sense used by *Trichogramma* to distinguish between parasitized and unparasitized hosts, *Proc. Roy. Soc. London*, **122**(B826), 57–75 (1937).

Severin, H. H. P., and H. C. Severin: Relative attractiveness of vegetable, animal, and petroleum oils for the Mediterranean fruit fly (*Ceratitis capitata* Wied.), *J. N. Y. Entomol. Soc.*, **22**(3), 240–248 (1914).

——, and ——: Behavior of the Mediterranean fruit fly (*Ceratitis capitata* Wied.) towards kerosene, *J. Animal Behavior*, **4**(3), 223–227 (1914a).

Sharma, H. N., and S. K. Sen: Oviposition in Culicidae, *Rep. Proc. 4th Entomol. Meeting Pusa*, 192–198 (1921).

Sharp, W. E.: *Melanophila acuminata* DeG. in Berkshire, *Entomol. Monthly Mag.*, **54**, 244–245 (1918).

Sihler, H.: Die Sinnesorgane an den Cerci der Insekten, *Zool. Jahrb. Anat.*, **45**(4), 519–580 (1924).

Thorpe, W. H., and F. G. W. Jones: Olfactory conditioning in a parasitic insect and its relation to the problem of host selection, *Proc. Roy. Soc. London*, **124** (B834), 56–81 (1937).

Totze, R.: Beiträge zur Sinnesphysiologie der Zecken, *Z. vergleich. Physiol.*, **19**(1), 110–161 (1933).

Townsend, C. H. T.: A record of results from rearings and dissections of Tachinidae, *U.S. Dep. Agr. Bur. Entomol., Tech. Ser.*, **12**(6), 95–118 (1908).

Travis, B. V.: Habits of the June beetle *Phyllophaga lanceolata* (Say) in Iowa, *J. Econ. Entomol.*, **32**(5), 690–693 (1939).

Ullyett, G. C.: Host selection by *Microplectron fuscipennis* Zett. (Chalcididae, Hymenoptera), *Proc. Roy. Soc. London*, **120**(B817), 253–291 (1936).

Valentine, J. M.: The olfactory sense of the adult meal-worm beetle *Tenebrio molitor* (Linn.), *J. Exp. Zool.*, **58**, 165–228 (1931).

Van Dyke, E. C.: Buprestid swarming, *Pan-Pacific Entomol.*, **3**(1), 41 (1926).

——: *Melanophila consputa* Lec., *Ibid.*, **4**(3), 113(1928).

Wieting, J. O. G., and W. M. Hoskins: The olfactory responses of flies in a new type of insect olfactometer, *J. Econ. Entomol.*, **32**(1), 24–29 (1939).

Williams, C. M.: Dry-cleaning fluid as an attractant for the kelp-fly, in "Laboratory Procedures in Studies of the Chemical Control of Insects," Washington, D.C., Am. Assoc. Adv. Sci., 1943, p. 174.

Wojtusiak, R. J.: Weitere Untersuchungen über die Raumorientierung bei Kohlweisslingsraupen, *Bull. Intern. l'acad. polon. sci., Classe sci. math. nat. B.*, *II*, 9/10, 631–655 (1930).

Worthley, H. N., and J. E. Nicholas: Tests with bait and light to trap codling moth, *J. Econ. Entomol.*, **30**(3), 417–422 (1937).

Chapter 3
Essential Oils, Resins, and Related Substances

Food Plant Selection. Nowhere is the rôle of attractants which guide insects to their proper food more complex and outstanding or the attractants more specific in action than in the lives of plant-feeding insects. This is especially true as it relates to those insects which feed on leaves. More work has been directed toward leaf-feeders than any others partly because of the great economic importance of this group, partly because of the accessibility of the insects and the comparative ease with which they may be studied.

Host specificity and the host preferences of phytophagous insects have long remained a challenging problem. Continued lack of progress in this field was due largely to a disinclination on the part of early investigators to realize that the relation between a phytophagous insect and its host is no less a chemical one than a botanical one. Moreover, the problem of plant selection was frequently confused with that of nutrition. Since the nutritive value of various angiospermous leaves is not markedly dissimilar, as evidenced by the fact that a given monophagous larva may by various experimental artifices be fed on a variety of plants, factors governing choice may be largely divorced from nutritional requirements. Essentially the problem of plant choice resolves itself into a study of attractants and repellents and vice versa.

For this study lepidopterous larvae are ideal experimental subjects. They are notoriously fastidious in their tastes; a more thorough knowledge of their feeding habits has been amassed than for any other single group of phytophagous insects; and as crop and shade tree pests they are always available in numbers.

Although the plant destined to nourish a larva during its several instars is initially selected by the ovipositing adult, the larvae themselves are capable of distinguishing certain plants. In many species this ability is a necessity since eggs are laid not on the food-plant but merely in its vicinity. This is especially true of some species of *Argynnis* and *Anartia*, the adults of which customarily oviposit on small bits of dead vegetation adjacent to food-plant seedlings. Also, not infrequently adults err in their choice of food-plant for oviposition. *Anartia jatrophae jamaicensis* Möschler has been observed at times ovipositing consistently on *Tradescantia* upon which the larvae subsequently refuse to feed. No specimens of *Lippia*, the usual food-plant, were growing in the immediate vicinity. A similar instance has been observed where the skipper *Poanes hobomok* Harr. laid 35 eggs on tansy, a plant with an unmistakable pungent odor. Larvae which emerged from these eggs would under no circumstances feed on tansy but developed rapidly when given their usual food, various grasses. Schwarz (1923) suggested that in some groups (e.g., *Catacola*) sense perception decreases with the onset of senility and that this accounts for adults ovipositing on plants not acceptable to the larvae. While this explanation may suffice in some cases, it is not adequate in others, notably that of *Poanes hobomok* where the adult was newly emerged.

The immediate factors influencing insects in their choice of food-plants are of two fundamental sorts—chemical and physical. The former are attractant or repellent in nature; the latter, usually repellent. Initially an insect is attracted to a plant by its odor; geotaxis and phototaxis merely direct the animal to the proper vicinity.

That odor might be one factor influencing food-plant choice by lepidopterous larvae was first suggested by Verschaffelt. This hypothesis was subsequently substantiated by several workers. While various investigators (Peterson, 1924; Morgan and Lyon, 1928; Morgan and Crumb, 1928; Gilmore and Milam, 1933; Moore, 1928; Raucourt and Trouvelot, 1936) delved into the chemical aspects of the problem on the assumption that olfaction was the primary sense involved, actual proof of this had not been forthcoming until the advent of McIndoo's (1926) work on coleopterous larvae and Dethier's (1937) on lepidopterous larvae.

McIndoo devised a piece of apparatus, termed an olfactometer, by means of which the reactions of insects to test odors could be indicated. Essentially this was a Y-tube through one arm of which passed ordinary air. Through the other arm passed a stream of air bearing an odor. By counting the number of insects in each arm at the end of the experiment one could judge whether under those conditions the test odor behaved as an attractant or a repellent. By this comparative method it was found that potato beetles are attracted by water extracts, aerated steam distillate, and unaltered odors from living plants. Here was the first experimental proof that plants attract insects by odor. Similar tests also indicated that tussock moth caterpillars (*Hemerocampa leucostigma* S. & A.) are slightly attracted to the odors of water extract and steam distillate of elm.

Three series of experiments designed to ascertain the nature of plant stimuli attractive to larvae of the milkweed butterfly (*Danais plexippus* L.) also emphasize the importance of odor in plant selection (Dethier, 1937). Series one, so-called "screen tests," was especially significant. These tests were so designed that only olfactory stimuli and hygrostimuli reached the larvae. Fresh leaves from various plants including milkweed were placed on the floor of a cage and covered with a wire screen which pressed them flat. Larvae released in the cage were then able to crawl about in close proximity to the leaves without being able to touch them. When the reactions of the larvae were observed and the paths plotted (Fig. 10), it was found that larvae passing over "foreign" leaves maintained straight paths while those passing over milkweed leaves described zigzag paths and exploratory movements of the head. Exploratory movements were most frequent whenever larvae passed above the

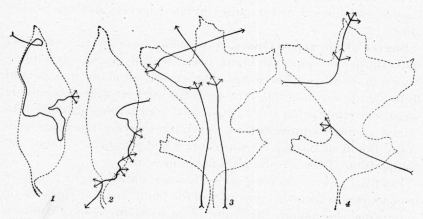

FIG. 10. (*1 and 2.*) Diagrams of typical paths of *Danais plexippus* larvae over their food plant (*Asclepias*) during a "screen test." (*3 and 4.*) Diagrams of typical paths of *D. plexippus* larvae over a foreign leaf (oak). The grouped arrows indicate locations where the larvae felt around in all directions. (From Dethier, *Biol. Bull.*, 1937.)

edge of a leaf. Fifty per cent of the time they turned back over the leaf.

Most significant of the experiments in this series were those in which the following leaves were placed under the screen: mullein (*Verbascum Thapsus* L.), oak (*Quercus ilicifolia*), plantain (*Plantago major* L.), mullein coated with milkweed latex, oak coated with milkweed latex, and plantain coated with milkweed latex. Plotted paths showed definitely that the caterpillars recognized the latter three leaves immediately.

If, as the above-mentioned experiments seemed to indicate, larvae recognized milkweed by some odor that it emitted, leaves coated with an odorless but disagreeable-tasting solution should have remained attractive. To test this supposition, milkweed leaves coated with solutions of sucrose, NaCl, or HCl were offered to larvae. Such treated leaves were immediately recognized and attempts made to consume them. No attempts were ever made to consume leaves coated with odorous substances. We may assume in the first instance that the leaf possessed the correct odor but the wrong taste. Finally, larvae were offered filter paper and certain other leaves coated with milkweed latex, as well as leaf sandwiches. Sandwiches were made by stripping the epidermis from both surfaces of a milkweed leaf and gluing it with latex to either side of a "foreign" leaf. Larvae endeavored to eat paper, treated leaves, and sandwiches alike.

Source of Attractive Odors. It is obvious from these experiments that odorous substances in plants constitute the attractants guiding insects to their food. The scents of practically all plants may be said to be due to essential oils. The term essential oil is an exceedingly broad one, embracing as it does all oily, odoriferous substances obtained almost exclusively from vegetable sources. These are generally liquid at ordinary temperatures and volatilize without decomposition. They are slightly soluble in water.

Essential oils, or volatile oils as they are sometimes called, have been known to occur in every part of the plant. In some cases the oil may occur throughout the entire plant as in the case of conifers; in others it is restricted to a certain part or organ of the plant. Frequently each different part of the plant may possess a distinct and totally unrelated oil. Thus, for example, the sassafras tree, *Sassafras variifolium* (Salisbury) O. Kuntze, contains in its roots an oil

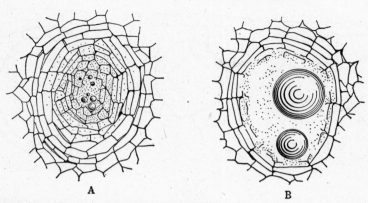

A   B

Fig. 11. Oil storage space of lysigenous origin in *Citrus vulgaris*. (A) Early stage. (B) Late stage showing advanced dissolution of secretory cells and formation of large oil droplets. (Redrawn after Tschirch, "Angewandte Pflanzenanatomie," Wien, Urban and Schwarzenberg, 1889.)

composed primarily of safrol, camphor, and eugenol while the principal constituent of oil from the leaves is the olefinic terpene, myrcene, plus citral, linalool, and geraniol. The Ceylon cinnamon tree, *Cinnamomum zeylanicum* Breyne, possesses at least three distinct oils each characterized by a different chemical, root oil by camphor, leaf oil by eugenol, and bark oil by cinnamic aldehyde.

Microscopically, essential oils are never present in a free state within the living cell. The oils may occur in the fibrovascular as well as the fundamental tissues. Receptacles may be either intact cells from which the protoplasm has disappeared or vascular structures originating by the gradual absorption of adjacent cell walls and the subsequent fusion of numerous cells into a single large vessel. Again the storage areas may be intercellular spaces formed by (1) decomposition of a number of adjacent cells (lysigenous origin, Fig. 11) or by (2) separation of adjacent cell walls without injury to the cells themselves (schizogenous origin, Fig. 12). Often a cavity may be of both lysigenous and schizogenous origin (Fig. 13). It is schizogenous in its inception and is enlarged by absorption of the cell walls. Such a vessel is termed schizolysigenous. Initially the oil may be secreted in various organs such as cells, hairs, vessels, etc. It may remain stored in the place of original secretion or may be transported to some other special storage place. Knowledge of the mechanism of actual secretion, however, is still in the hypothetical stage. The first

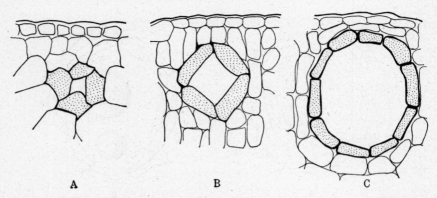

FIG. 12. Semidiagrammatic representation of the formation of schizogenous storage spaces. (A) Initial separation of four mother cells. (B) Advanced stage. (C) Fully formed storage space.

but less generally accepted theory is that of Tschirch (see Parry, 1918).

According to Tschirch the external portions of the cell membranes which border on the storage vessel become mucilaginous and form the first products of the transformation of the cell substance into the essential oil which then appears in the vessel as tiny oil droplets. This conversion proceeds rapidly until the fully developed vessel is completely surrounded by the secreting cells whose membranes on the side bordering the vessel are jellified, forming the resinogenous layer. The secretion has not diffused through the actual cell

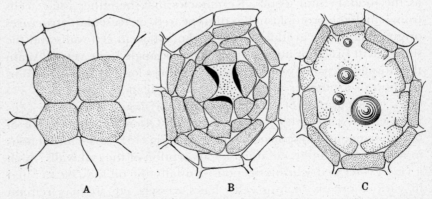

FIG. 13. Schizolysigenous formation of an oil storage space in *Citrus Aurantium*. (A) Initial separation of four mother cells. (B) Secretion of oil and increase in number of secretory cells. (C) Dissolution of secretory cells and formation of large oil droplets. (Redrawn after Tschirch, "Angewandte Pflanzenanatomie," Wien, Urban and Schwarzenberg, 1889.)

membranes as an essential oil but as an intermediate product. The actual genesis of the oil is in the resinogenous layer.

There is considerable evidence, however, that the oil does pass through the membrane and little evidence that such a thing as a resinogenous layer actually does exist. Tschirch postulated this layer because he did not believe that the oil could permeate water-saturated membranes. He overlooked the fact that essential oils are slightly soluble in water.

Tunmann concluded that secreted matter is only found outside the glandular cells and is divided from the cell plasma by a visible wall of cellulose. He claimed, therefore, to have discovered the resinogenous layer of Tschirch. Moreover, he was able to determine various typical forms of this layer, namely: the rod-type (*Viola, Fraxinus, Alnus*), the vacuola-type (*Salvia, Hyssopus*), and the mesh-type (*Rhododendron, Azalea*).

In the case of all the persistent glands of the Labiatae, Pelargoniae, Compositae, etc., all of which possess a strong cuticle, continuous volatilization of the essential oil occurs throughout the lifetime of the plant. In the course of this process the chemical composition of the essential oil must undergo some modification, but it does not reach a demonstrable process of resinification because new volatile portions are continually being formed. Only in autumn when the period of growth is reaching its end does the formation of volatile constituents cease and the remainder of the oil resinify. Thus autumn leaves, in lieu of the usual, almost colorless, highly refractory essential oil, contain dark yellow, partly crystalline, partly amorphous, somewhat sparingly soluble lumps of resin.

Charabot and Laloue (see Parry), by laborious experimentation, deduced the following facts concerning oil secretion:

The odorous compounds first appear in the green organs of the plant whilst still young. They continue to be formed and to accumulate till the commencement of flowering, but the process becomes slower as the flowering advances. The essential oil passes from the leaves to the stems and thence to the inflorescence, obeying the ordinary laws of diffusion. Part of it, entering into solution, passes into the stem by osmosis. Arriving here, and finding the medium already saturated with similar products, precipitation takes place, the remaining soluble portion continuing to diffuse, entering the organs where it is consumed, especially the inflorescence. Whilst fecundation is taking place a certain amount of the essential oil is consumed in the inflorescence. It is possible, and even probable, that at the same time the green organs are producing further quantities of the

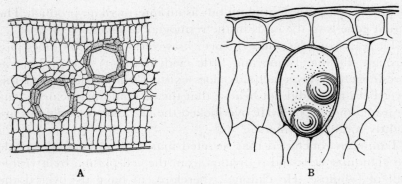

FIG. 14. (A) Oil storage spaces in the leaf of *Melaleuca minor*. (B) Oil cell from the leaf of *Sassafras officinale*. (Redrawn after Tschirch, "Angewandte Pflanzenanatomie," Wien, Urban and Schwarzenberg, 1889.)

essential oil, but all that can be said with certainty is that a net loss in essential oil occurs when the flower accomplishes its sexual function.

If all this is true, the essential oil circulates in an aqueous medium and the postulation of a resinogenous layer is unnecessary and unacceptable.

In leaves which are fast of their smell, such as bay, myrtle, etc., the globule of oil is enclosed in a relatively thick capsule embedded

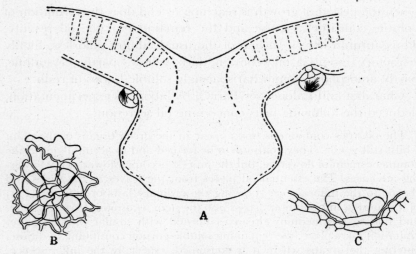

FIG. 15. (A) Transverse section through a leaf of *Mentha piperata* showing two typical labiate oil glands. Note menthol crystals. (B) Enlarged surface view of a labiate oil gland. (C) Side view of same. (Redrawn after Tschirch, "Angewandte Pflanzenanatomie," Wien, Urban and Schwarzenberg, 1889.)

in the leaf (Fig. 14). In thyme and many other labiates, which give off their scent at a touch or under the heat of the sun, the oil is stored on the surface of the leaf in easily broken flask- or goblet-shaped cells. In the genus *Cistus* the oil is secreted onto the surface of the leaf and mixed with a sticky resin which retards evaporation. In rue the oil is contained in a cavity beneath the surface of the leaf which is roofed over by a single layer of cells and pierced in the center by a narrow opening (Fig. 16). The cells of the roof are alive. They can swell up and become turgid. In so doing they bend inward, thus enlarging the orifice and driving the oil to the surface of

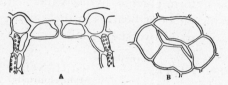

FIG. 16. Side (A) and surface (B) views of a specialized oil cavity in *Ruta grave-olens*. (Redrawn after Haberlandt, "Physiologische Pflanzen-anatomie," Leipzig, W. Engelmann, 1904.)

the leaf. The oil of the flower is contained in the epidermal cells of the upper side of the petals, sepals, and bracts.

With regard to the physiological status of the essential oil in the plant little can be said with certainty. The generally accepted view is that essential oils represent waste products of the plants. Supporting this contention is the presence of several substances such as phenylethyl alcohol, indole, trimethylamine, and ethyl butyrate in essential oils, substances which are known to occur among the products of decomposition of protein material, either as end-products or by the recombination of these end-products with one another. The oil has been postulated as arising from starch, cellulose, chlorophyll, and even tannin.

Be that as it may, most essential oils appear to be directly evolved from terpenes or from glucosides (see pp. 52 and 56). In the latter case the glucoside is acted upon by soluble ferments such as diastases, enzymes, and similar bodies, which are hydrolytic agents, in order to produce the constituents of the essential oil. Usually an essential oil arising from glucosidal decomposition is nonterpenic in character. The specific ferments able to decompose particular glucosides are usually found in the plant containing the glucoside but separated from the latter by being enclosed in special cells. Decomposition may be of two kinds. First, hydrolysis may occur within the plant itself so that the essential oil is actually a product of the metabolic processes. Second, the decomposition may follow as a result of artificial crushing of the plant such as would occur if the plant were eaten by some animal. A case which will illustrate the latter type of decomposition is that of certain of the Rosaceae where amygdalin ($C_{20}H_{27}NO_{11}$) is acted upon by a ferment, emulsin (which is actually two enzymes, amygdalase and prunase, and is rarely, if ever, in contact with amygdalin in the plant tissues), in the presence of water to form glucose, hydrocyanic acid, and benzaldehyde, the characteristic odor-bearers in these plants. The reaction, which probably occurs in two stages, follows:

$$C_{20}H_{27}O_{11}N + H_2O \xrightarrow{\text{Amygdalase}} \underset{\substack{\text{Mandelonitrile} \\ \text{glucoside}}}{C_{14}H_{17}NO_6} + \underset{\text{Glucose}}{C_6H_{12}O_6}$$

$$H_2O + C_{14}H_{17}NO_6 \xrightarrow{\text{Prunase}} \underset{\text{Benzaldehyde}}{C_7H_6O} + HCN + C_6H_{12}O_6$$

To complicate the situation, essential oils are by no means constant within a species. Many factors, including climate and soil variations, may affect the composition of an oil. Ordinarily, for example, oil from *Thymus vulgaris* L. consists largely of thymol [isopropyl-meta-cresol, $C_6H_3CH_3OHC_3H_7$], but this may be more or less completely replaced by its isomer carvacrol [isopropyl-ortho-cresol, $(CH_3)_2CHC_6H_3CH_3OH(4,1,2)$]. Again, oil from wild lavender flowers, *Lavandula vera* D. C., in France owes its odor primarily to linalool [$C_{10}H_{18}O$] and its esters in addition to geraniol, while oil from cultivated English plants consists largely of cineol which has an entirely different odor.

As has already been mentioned, essential oils are mixtures of many chemical substances. The attar of flowers, or flower oils, are all similar in that every one contains carbon and hydrogen. Many contain oxygen, and some, sulfur and nitrogen. The chief chemical constituents of essential oils are: esters (seldom occurring singly), alcohols, acids (not often found free), aldehydes (usually common in minute amounts), ketones (violet scent is always due to a ketone—ionone in violets and irone in orris root—and the smell of rue is due to methylnonylketone), phenols, phenol esters, lactones, terpenes (occurring in practically all essential oils not of glucosidal origin), benzene compounds, and nitrogen compounds (trimethylamine and propylamine give an unpleasant, fishy, ammoniacal smell to flowers of hawthorn and leaves of *Chenopodium vulvaria* and Dog's Mercury). They may be aromatic or aliphatic compounds, saturated or unsaturated.

So numerous and complex are the compounds which go to make up essential oils that attempts at satisfactory classifications are seldom successful. They are certainly arbitrary. Gortner (1938) lists eleven categories representing typical examples of these compounds.

1. Terpenes: hydrocarbons of the general formula $C_{10}H_{16}$.

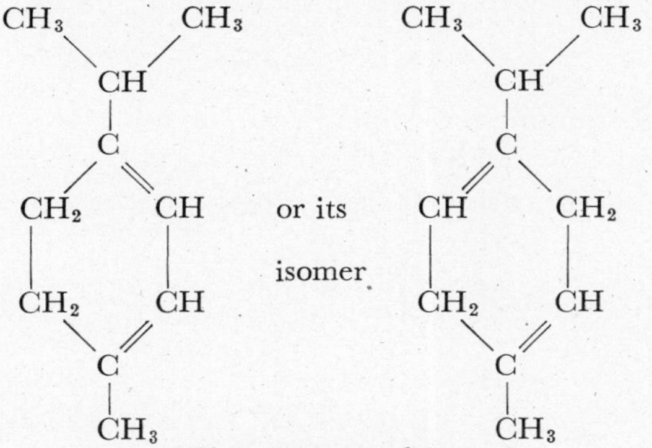

The terpene nucleus

Examples of this type of compound are pinene, limonene, camphene. Terpenes and members of the fatty acid series predominate in lower plants while the higher plants are characterized more by aromatic sulfur and nitrogen compounds.

2. Alcohols and ketones: camphor series.

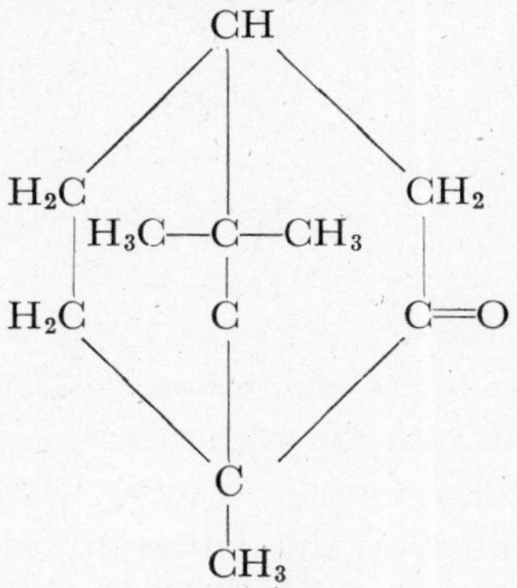

Camphor

Typical of this series are cineol, menthol, borneol.

3. Geraniol and citronella group: no closed ring.

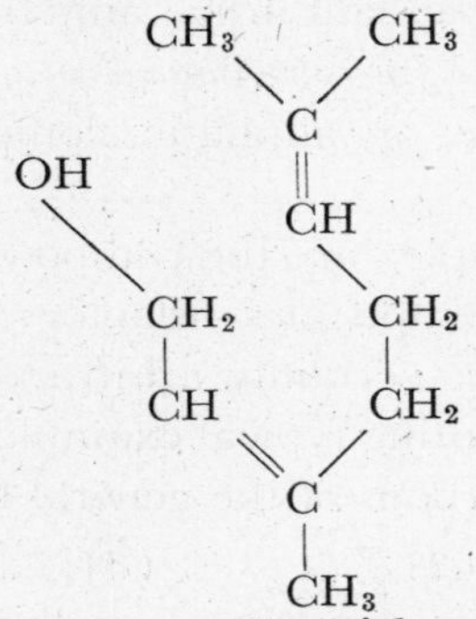

$$CH_3 \quad CH_3$$
$$\backslash / $$
$$C$$
$$OH \quad \| $$
$$\backslash \quad CH$$
$$CH_2 \quad CH_2$$
$$| \qquad | $$
$$CH \quad CH_2$$
$$\backslash / $$
$$C$$
$$| $$
$$CH_3$$

Geraniol

One of the commonest of this group in addition to geraniol is citronellol which is found in lemon grass and citrus fruits.

4. Benzene hydrocarbons: exclusive of terpenes.

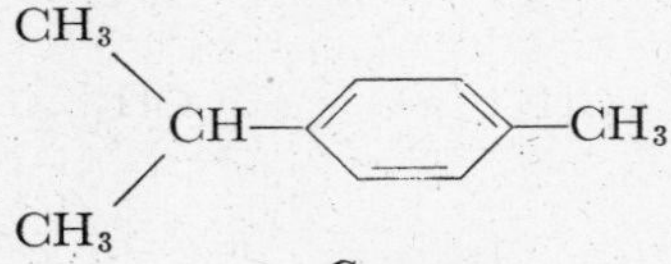

$$CH_3$$
$$\backslash$$
$$CH-\bigcirc-CH_3$$
$$/$$
$$CH_3$$

Cymene

These compounds are relatively common as, for example, cymene in caraway oil.

5. Phenols: This widely distributed group includes eugenol, thymol, carvacrol, and carvone.

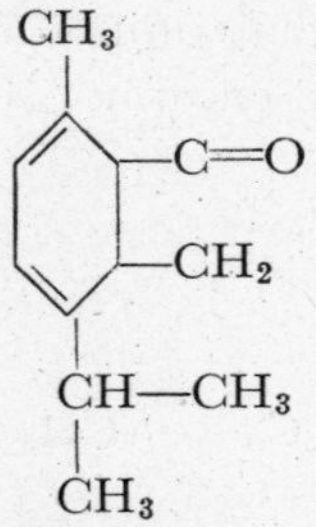

$$CH_3$$
$$|$$
$$-C=O$$
$$-CH_2$$
$$CH-CH_3$$
$$|$$
$$CH_3$$

Carvone

6. Acids and esters: These are common.

7. Aliphatic alcohols also are common.

8. Aliphatic aldehydes occur as traces.

9. Sulfides: Compounds of this group are especially common in the family Cruciferae. Two of the better known ones are allyl sul-

fide, $CH_2\!\!=\!\!CH\!\!-\!\!CH_2\!\!-\!\!S\!\!-\!\!CH_2\!\!-\!\!CH\!\!=\!\!CH_2$, and allyl thiocyanate, $CH_2\!\!=\!\!CH\!\!-\!\!CH_2\!\!-\!\!CNS$.

10. A few rare paraffin hydrocarbons are found in plants.

11. Organic bases (e.g., indole) also occur in small amounts.

In this huge assemblage of compounds are to be found the individual odorous substances which constitute the specific attractants guiding each phytophagous insect to its preferred food. The nicety of the relation between insects and plant chemicals, the manner in which these relations are coördinated, and the procedures followed in investigating them may be appreciated from the following studies which represent the extent of this field to date.

Mustard Oils and Related Substances. Certain species of *Pieris*, notably the greater and lesser cabbage butterflies (*P. brassicae* L. and *P. rapae* L.) in the larval stage ordinarily feed on plants of the family Cruciferae. Cultivated species are especially favored as food, but as far as is known all species may be attacked. The following representative species have been tested:

> *Cochlearia Armoracia* L.
> *Sisymbrium officinale* (L.) Scop.
> *S. strictissimum* L.
> *Sinapis arvensis* L.*
> *Brassica oleracea* L.
> *Crambe cordifolia* Stev.
> *Barbarea vulgaris* R. Br.
> *Cardamine hirsuta* L.
> *Capsella Bursa Pastoris* (L.) Mich.*
> *Aubrietia deltoidea* (L.) DC.*
> *Arabis alpina* L.
> *Erysimum Perofskianum* Fisch. et Mey.*
> *Alyssum saxatile* L.
> *Hesperis matronalis* L.
> *Bunias orientalis* L.

Those marked with an asterisk are eaten only slightly. Verschaffelt suspected that they might possess distasteful subsidiary constituents. That this might be the case is strongly indicated by similar phenomena associated with the feeding of *Papilio* larvae on Umbelliferae (see p. 64). Feeding by larvae of cabbage butterflies is not restricted to Cruciferae, however, for species of *Tropaeolum* and *Reseda* also are attacked. Acceptance of these two genera supplied the key to the solution of the question as to which chemical constituents of a plant determine the choice of *Pieris* larvae since they were found to contain essential oils similar to those so characteristic of Cruciferae. These are known collectively as mustard oils.

The term mustard oil embraces a group of substances consisting largely of sulfur compounds, especially alkyl isothiocyanates. These are various esters of isothiocyanic acid, HCNS, and are formed by hydrolysis of certain glucosides. In seeds of some species of *Brassica*, especially *B. nigra* Koch, the predominating compound is allyl isothiocyanate, CH_2=CH—CH_2—NCS. This compound results from the action of an enzyme, myrosin, on the glucoside, sinigrin (potassium myronate), as illustrated by the following equation:

$$\underset{\text{Sinigrin}}{C_{10}H_{16}NS_2O_9K} + H_2O \xrightarrow{\text{Myrosin}} \underset{\substack{\text{Allyl iso-}\\\text{thiocyanate}}}{C_3H_5NCS} + \underset{\text{Dextrose}}{C_6H_{12}O_6} + \underset{\substack{\text{Potassium}\\\text{hydrogen}\\\text{sulfate}}}{KHSO_4}$$

By similar reactions different alkyl isothiocyanates are formed in other species of Cruciferae: crotonyl isothiocyanate, CH_2=CH—CH_2=CH_2=NCS, from the seeds of *Brassica napus* L. and *B. campestris* L., isothiocyanate of secondary butyl alcohol, CH_3—CH_2—(CH_3)—CH—NCS, from *Cochlearia officinalis* L. In the family Resedaceae are still other oils. Characteristic of *Reseda odorata* L. is phenylethyl isothiocyanate, C_6H_5—C_2H_4NCS.

Mustard oils are more or less generally distributed also in the families Moringaceae, Limnanthaceae, Caricaceae, Resedaceae, Tropaeolaceae, and Capparidaceae. Of the first three families no species were available to Verschaffelt for experimentation. *Reseda lutea* L., *R. luteola* L., *R. alba* L., and *R. virgata* Boiss. and Reut. belonging to the family Resedaceae are eaten; *Tropaeolum majus* L. and *T. peregrinum* L. (Tropaeolaceae) and *Capparis spinosa* L., *Steriphoma paradoxum* Endl., and *Cleome spinosa* L. (Capparidaceae) also are readily eaten. As a control, leaves from a series of plants chosen at random were offered to both species of *Pieris*. A few, especially *Lathyrus sylvestris* L. and *L. latifolius* L., were eaten slightly. Since mustard oils are not known to occur in these species it was concluded that plants devoid of these compounds may be attacked slightly and sporadically. The correctness of this conclusion is suggested by comparable occurrences in experiments with other monophagous and oligophagous larvae.

To show beyond all manner of doubt that *Pieris* larvae are truly attracted by mustard oils, leaves of unacceptable plants—wheat, maize, starch, and filter paper—were smeared with juices and pastes of an acceptable plant, *Bunias orientalis*. With the exception of *Salvia officinalis*, *Prunus Laurocerasus*, and *Menyanthes trifoliata*, all so treated were eaten. The exceptions probably owed their immunity to attack to the presence of strong distasteful constituents.

Ordinarily, acceptable plants when intact contain unsplit glucosides, but the question remains whether the presence of a minute quantity of free mustard oil is not the reason why caterpillars are attracted. In some crucifers (e.g., *Bunias orientalis*) a faint odor of mustard oil can be perceived in the unbruised leaf. It is apparent that larvae are made aware of the nature of a plant by its odor since they seldom taste other species.

The positive reactions of larvae to solutions of pure sinigrin suggests that the glucoside as such might be attractive though there is a possibility that the glucoside is hydrolyzed by saliva and thus is attractive only as an oil.

It is clear also that larvae are attracted by the whole group of related substances comprehended under the term mustard oils. Experiments should be undertaken to determine exactly how wide a range of substances actually is attractive, whether the phenomenon is restricted to thiocyanates or includes other cyanates, whether attraction is characteristic only of those compounds containing both nitrogen and sulfur or of some compounds containing sulfur alone. That certain sulfur compounds lacking a nitrogen atom may be attractive is suggested by the readiness with which larvae may eat species of *Allium*, especially *A. cepa* L., *A. Porrum* L., *A. azureum* Ledeb., when the occasion is presented. The compounds to which species of *Allium* owe their odor were at one time thought to be alkyl sulfides, especially allyl sulfide, $(CH_2{=}CH{-}CH_2)_2S$. It is now suspected that much more complex sulfur compounds are involved; nevertheless, comprehension of the nature of reactions of larvae to alkyl sulfides would be enlightening.

The reactions of another insect, the adult onion maggot, *Hylemyia antique* Meig. (Anthomyiidae), to mustard oil, allyl sulfide, allyl iodide, allyl bromide, allyl alcohol, propyl alcohol, ethyl alcohol, and butyl alcohol have been tested (Peterson, 1924). Of these substances, mustard oil and allyl sulfide are slightly attractive; other allyl compounds, not so much so. The alcohols also are slightly attractive. Since the attractiveness of test baits was enhanced by the addition of yeast to promote fermentation, it would seem that at the time of testing adults were conditioned to, and reacting to, a feeding-type attractant rather than to sulfur compounds which are obviously ovipositional-type attractants to the adults and feeding-type attractants only to the larvae. From our point of view additional conclusive tests could be run with the larvae. As far as they go the tests indicate that attraction is not due to the allyl part of the compound. It is interesting to speculate on the type of reaction that might be provoked by the presentation of isothiocyanates to these larvae. Some work on the reaction of typical crucifer-feeding insects and *Allium*-feeders to the different odorous sulfur compounds of plant origin would be welcomed. Reactions especially to the following should be recorded: dimethyl sulfide, $(CH_3)_2S$, a foul-smelling oil occurring in small amounts in peppermint and geranium; vinyl sulfide, $CH_2{=}CH_2S$, an evil-smelling compound in *Allium ursinum;* diallyl sulfide, $(C_3H_5)S{-}S(C_3H_5)$, the essence of garlic; allyl-propyl disulfide, $(C_3H_5)S{-}S(C_3H_7)$; secondary butyl isothiocyanate, $CH_3{-}CH_2{-}CH(CH_3){-}N{=}C{=}S$, possessing the odor of sulfur; benzyl isothiocyanate, $C_6H_5{-}CH_2{-}N{=}C{=}S$, which imparts a characteristic odor to *Tropaeolum;* phenylethyl isothiocyanate, $C_6H_5(C_2H_4)N{=}C{=}S$, found in oil of reseda root, nasturtium, and some species of *Brassica;* oxybenzyl isothiocyanate, $C_6H_4{-}OH{-}CH_2NCS$, from white mustard.

Benzaldehyde and Hydrocyanic Acid. As mustard oils are formed by the action of an enzyme upon a glucoside so also is formed oil of bitter almonds, the chief constituents of which are benzaldehyde and hydrocyanic acid. The equation illustrating this hydrolysis has been given on p. 48. The two odorous compounds are rather widely distributed throughout the plant kingdom. According to Greshoff (1906) there are approximately 84 genera of flowering plants known to contain hydrocyanic acid, but only 43 of these contain benzaldehyde also. Together the two compounds are especially characteristic of three tribes of Rosaceae, namely, Spiraeae, Pomeae, and Pruneae.

Of the numerous insects more or less restricted in their feeding habits to rosaceous plants, only two have been the subject of extensive experimentation along these lines, larvae of the leaf wasp

Periophorus padi L., and the American tent caterpillar, *Malacosoma americana* Fabr.

Verschaffelt tested the feeding reactions of the former to a selection of Rosaceae of which the following are preferred species.

    Cotoneaster tomentosa Lindl.

    Mespilus germanica L.

    Amelanchier vulgaris Monch.

    Crataegus Oxycantha L.

    C. Pyracantha L.

    Cydonia vulgaris Pers.

    C. japonica Pers.

    Sorbus Aucuparia L.

    S. americana Pursh.

    Prunus Persica Sieb. et Zucc.

    P. avium L.

    P. cerasus L.

    P. Laurocerasus L.

    P. Padus L.

Chemical analysis reveals that the above listed species contain the glucoside amygdalin in such quantity that the odor of benzaldehyde and HCN is exceptionally pronounced when the leaves are crushed. Of the Rosaceae refused, some contain no amygdalin or such low concentrations (*Spiraea*, *Kerria*, and *Pirus*) that the larvae do not react positively.

While experiments involving the smearing of nonacceptable leaves with solutions of amygdalin were by no means so uniformly successful as those with solutions of sinigrin in experiments with *Pieris*, they do indicate that the chief sources of attraction are amygdalin and its fission products.

The tent caterpillar also more or less consistently restricts its feeding to members of the Rosaceae. Preferred foods are species of wild cherry, notably *Prunus virginiana* L., and apple. Plum, peach, thorn, pear, rose, and other Rosaceae are less frequently eaten while under special and rare conditions beech, willow, witch hazel, elm, oak, and poplar trees may be defoliated. A glance reveals that the preferred food plants belong to the tribes Pruneae and Pomeae where a high concentration of odorous chemicals is to be expected. In wild cherry the concentration of HCN is so high that cattle ingesting wilted leaves are severely poisoned. Yet, if the juice from wild cherry leaves, expressed by crushing, is sprayed on filter paper; larvae readily eat the treated paper. The two chief constituents of this juice, HCN and benzaldehyde, fail to attract when sprayed on paper separately; however, an emulsion of approximately 1 cc. of a mixture of equal parts of the two compounds and 1 to 2 liters of water acts as a mild attractant.

Even though natural food is preferred to any such emulsion, the positive responses of larvae indicate the nature of the attractants at work. No responses to the glucoside itself have been observed. Apparently it is without effect.

The exact significance of glucosides as possible attractants is confusing. Verschaffelt claimed to have observed positive responses by larvae of *Pieris* to sinigrin. On the other hand, no one has been able to induce larvae of Rosaceae-feeding insects to respond to amygdalin. There is reason to expect that the specific glucosides would elicit similar responses in each instance. Just what are the characteristics of this kind of compound? Glucosides are composed of a nonsugar group, usually of an aromatic type, attached to carbon 1 of a sugar. Most are crystalline, colorless, bitter, soluble in water and alcohol, and *without a pronounced odor*. The distinguishing odors of plants containing them must therefore be due primarily to fission products even though it is claimed that these seldom occur free or uncombined in plants. Even HCN, which is fairly common in higher plants, exists in a combined state with a cyanogenetic or cyanophoric glucoside. Uncombined it is extremely rare. Until further investigation is conducted we must assume that in the instances discussed above fission products rather than the glucosides themselves are the prime attractants and that these operate in plant selection even though they supposedly do not preëxist in the plant. The suggestion by Verschaffelt that the glucoside might be hydrolyzed by the saliva of the insects and thus attract fails to explain the initial attraction of the insects to the plant—a phenomenon proved repeatedly to be regulated by odor.

Naturally occurring glucosides, though widely distributed in nature, occur only in small amounts. For further information a table of certain of these, grouped according to the nature of the nonsugar part of the molecule, is included in Table 3.

Oils from Umbelliferae. As the foregoing experiments so aptly indicate, a successful solution to problems bearing on plant attractants lies not only in being able to isolate the active odoriferous principle but also in being able to identify it and proffer it to the insect in satisfactory concentration. It is relatively simple to discover that the odor of a given plant attracts certain insects. It is difficult, on the other hand, to ascertain exactly what compound or compounds are acting as the attractant.

The most satisfactory procedure followed thus far has been that of spraying the test substance on filter paper, disks of pith, or thin sheets of paraffin. It is always desirable to work with low concentrations since nearly all chemicals which are attractive at low concentrations act as repellents at high concentrations. This was demonstrated very clearly by Hoskins and Craig (1934). One way of

Table 3

REPRESENTATIVE NATURALLY OCCURRING GLUCOSIDES*

Name	Source
Coniferin	Bark of fir
Populin	Bark of poplar
Salicin	Bark of willow
Prunasin	Twigs of *Prunus padus*
Aesculin	Bark of horsechestnut
Apiin	Leaves of some Umbelliferae
Delphinin	Flowers of larkspur
Pelargonin	Flowers of geranium
Digilanide	Leaves of foxglove

* Reprinted by permission from "Outlines of Biochemistry," by Ross A. Gortner, published by John Wiley & Sons, Inc., 1938.

attaining a low concentration is through the simple expedient of allowing filter paper to stand exposed to the air for several days following treatment. The resultant volatilization effectively lowers concentration. It is also desirable to moisten the paper as this more nearly simulates natural conditions, for, as Ramsay, Butler and Sang (1938) have shown, there is naturally a marked humidity gradient at the surface of transpiring leaves (Fig. 17). The amount of filter paper or paraffin consumed serves as an index of the attractiveness of the chemical in question. Under ideal experimental conditions larvae may be induced to consume quantities of treated material though, needless to say, the nutritive value is nil. At times it is more satisfactory to apply the attractant to some mild leaf which

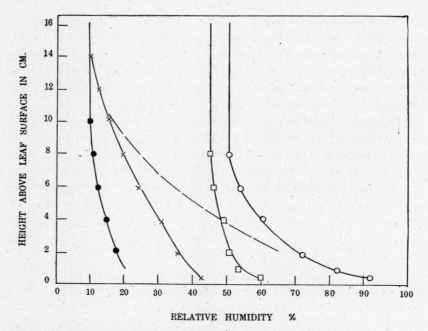

FIG. 17. Humidity gradient at the surface of transpiring leaves. Solid circles and crosses, *Rumex hydrolapathum;* squares and open circles, tulip leaves; broken line, wet filter paper. (Redrawn after Ramsay, Butler, and Sang, *J. Exp. Biol.,* 1938.)

has no pronounced attractant or repellent features of its own. Humidity is thus controlled, and the final results are not obscured by the insect's inability to masticate fibers of paper.

Studies embodying this technique and the pattern of earlier work have revealed the chemical factors which determine the choice of food-plants by the larvae of swallowtail butterflies of the genus *Papilio* (Dethier, 1941).

In current literature the larvae of *Papilio ajax* L. are listed as feeding upon various Umbelliferae. In order to verify this statement all members of the Umbelliferae available at the time of testing were fed to larvae. As a result it was found that the following umbelliferous plants may be eaten to a varying degree:

Pteroselinum sativum Hoffm.

Cicuta maculata L.

Conium maculatum L.

Sium circutaefolium Schrank

Daucus Carota L.

Osmorrhiza longistylis (Torr.) DC.

Carum Carvi L.

Pastinaca sativa L.

Anethum graveolens L.

Apium graveolens L.

To this list may be added the following umbelliferous plants reliably reported in the literature:

Cicuta virosa L.

C. bulbifera L.

Hydrocotyle umbellata L.

Foeniculum vulgare Hill.

Angelica atropurpurea L.

Echinophora spinosa L.

Arracac xanthorrhiza Baner.

Oxypolis filiformis (Walt.) Britton.

The absolute refusal of many larvae to eat *Hydrocotyle americana*, a plant that has been reported only once as food, suggests that all species of Umbelliferae may not be acceptable. It is quite important that such gaps in our knowledge be filled. Under the circumstances, however, it is necessary to base tentative hypotheses upon the characters of the species known to serve as food plants.

The Umbelliferae are characterized by a variety of odors which are said to have a family resemblance. That odors may possess a family resemblance and at the same time retain their specificity is a recognized fact (Delange, 1930; Dumont, 1928). To the human sense of smell the common odors of the Umbelliferae are seven, to wit: carrot, caraway, anise, coriander, celery, ferula, and angelica. Because of lacunae in our present knowledge it is not yet possible to consider other of the less common odors in this plant family.

As is usual among plants, these odors originate with the essential oils. The substance giving rise to the odor of carrot is not known. Caraway odor arises from carvone; anise odor from methyl chavicol, anethole, anise ketone, and anise aldehyde; coriander odor from coriandrol; celery odor from sedanolid; ferula odor from disulfides, especially $C_{11}H_{20}S_2$ (Table 4). On the basis of the current food plant list for *Papilio ajax*, ferula odor may be omitted from further discussions. The question then arises as to whether *P. ajax* is attracted by six different odors or confuses the six as one. There is reason to believe, as pointed out below, that the odors are not confused.

The oils, and in some cases the chemicals, from which arise the odors of carrot, caraway, anise, coriander, and celery, were procured in pure form. Although the odor of a plant is the result of a mixture and blending of incredibly minute quantities of chemicals, there is usually one chemical fundamentally responsible for the odor. The odor of this chemical resembles that of the plant suffi-

Table 4

DISTRIBUTION OF IMPORTANT ESSENTIAL OILS IN THE UMBELLIFERAE

Methyl Chavicol	Anethole	Carvone	Coriandrol	Sedanolid
Anthriscus Cerefolium Hoffm. Foeniculum piperitum Sweet. Pimpinella Anisum L.	Osmorrhiza longistylis (Torr.) DC. Foeniculum vulgare Hill. F. piperitum Sweet. Anthricus Cerefolium Hoffm. Myrrhis odorata Scop. Pimpinella Anisum L. Anethum	Carum Carvi L. Anethum graveolens L. Pseucedanum graveolens Benth. & Hook.	Coriandrum sativum	Apium graveolens L. Selinum

ciently closely in most cases to attract the insect; therefore, pieces of filter paper were soaked in the different pure oils, some in methyl chavicol, some in carvone, some in carrot oil, some in coriandrol, and some in celery oil. Although all of the *P. ajax* larvae tested attacked the filter paper thus treated, indicating that the oils serve as attractants, most of them did not consume any appreciable amounts because of the toughness of the fibers.

On the basis of these experiments it followed that *P. ajax* larvae should be attracted by the same substances when present in other plants provided that the odor due to the attracting compounds was not masked by some other more odoriferous constituent. Certain other plants in addition to the Umbelliferae are known to contain carvone, methyl chavicol, anethole, anisic acid, anisic aldehyde, and coriandrol. These are listed in Table 5. Theoretically, the larvae should attack the leaves of these species if no masking odor is present. As a matter of fact they do eat *Dictamnus Fraxinella, Cosmos,* certain species of *Solidago*, and *Artemisia dracunculoides*, a species closely related to *A. Dracunculus*.

Although larvae should be attracted to these plants, it does not follow that they should eat them for other factors of secondary importance, such as leaf thickness, pubescence, etc., come into play and prevent the utilization of the leaf as food.

When one deals with attracting chemicals in pure form, preferences on the part of larvae become quite obvious. Of all the oils tested above, those with an anise odor proved most attractive to larvae. The caterpillars with which the tests were made had been taken from cultivated carrots. Filter paper treated with methyl chavicol was preferred to fresh carrot leaves. Carrot was preferred to carvone, carvone to oil of coriander. There can be little doubt as to the ability of the larvae to distinguish the different odors. This would seem to indicate, therefore, that an oligophagous species such as *P. ajax* may be conditioned to more than one odor. It would likewise appear that the olfactory sense in these larvae is capable of discriminating among closely related odors. In this respect it simulates the human sense of smell.

Larvae of *P. ajax* have been found feeding also on *Ruta graveolens* L. and *R. patavina* L. the pungent rue odor of which emanates from methylnonylketone present in the leaves. On the basis of the same line of reasoning as above it follows that other plants containing this chemical should be attractive to larvae. The following plants are known to contain methylnonylketone: *Citrus Medica* L., *Ruta montana* Mill., *R. graveolens* L., *R. chalepensis* DC., *Zanthoxylum ailanthoides* Sieb. et Zucc., *Z. senegalense* DC., *Empleurum serrulatum* Eckl. et Z., and *Pilocarpus Jaborandi*, all in the family Rutaceae, and *Litsea odorifera* in the family Lauraceae. *P. ajax* larvae had never been reported from any of these plants. Larvae confronted with *Ptelea trifoliata* L., which smelled faintly like rue, endeavored to eat it but proceeded with difficulty because of the leathery quality of the leaves. *Z. americanum* Mill., which possesses an orange or lemonlike odor plus a very faint suggestion of rue, also was sampled by the larvae which made repeated attacks on the leaves only to be repulsed by their extreme toughness. *Z. simulans* Hanc. possesses a carrotlike odor. It was readily attacked by larvae but was protected by the toughness of its leaves. *Z. schinifolium* Sieb. et Zucc., having a pronounced orange and lemon odor, was not eaten. In *Citrus Medica* the

Table 5

DISTRIBUTION OF OILS ATTRACTIVE TO *P. ajax* IN PLANT FAMILES OTHER THAN UMBELLIFERAE

Methyl Chavicol	Anethole	Anisic Aldehyde	Carvone	Anisic Acid
Pinaceae	Magnoliaceae	Rutaceae	Pinaceae	Loganiaceae
Pinus palustrus Mill.	Illicium anisatum Gaertn.	Barosma venustum Eckl. & Z.	Pinites succinifer Göpp.	Buddleia variabilis Hemsl.
P. contorta Dougl.	I. verum Hook.	Pelea madagascarica Baill.	Taxodium distichum Rich.	Betulaceae
P. Sabiniana Dougl.	Magnolia Kobus Kaempf.	Leguminosae	Libocedrus decurrens Torr.	Betula alba L.
P. Jeffreyi Murr.	Lauraceae	Acacia Farnesiana Willd.	Scitamineae	Ranunculaceae
Gramineae	Persea gratissima Gaertn.	A. Cavenia Hook.	Curcuma longa L.	Aconitum Napellus L.
Andropogon Nardus Rendl.	Burseraceae	Cruciferae	Verbenaceae	
Magnoliaceae	Canarium luzonicum Gray.	Cheiranthus Cheiri L.	Lippia adoensis Hochst.	
Illicium anisatum Gaertn.	Chenopodiaceae	Orchidaceae	Lauraceae	
I. verum Hook.	Roubieva multifida Moq.	Vanilla planifolia Andr.	Litsea sericea Bl.	
Magnolia Kobus Kaempf.	Piperaceae		Labiatae	
Lauraceae	Piper peltatum L.		Mentha verticillata var. strabala Briq.	
Laurus nobilis L.	Rutaceae		M. velutina Lej.	
Persea gratissima Gaertn.	Barosma serratifolia Willd.		M. aquatica L.	
Burseraceae	B. venustum Eckl. & Z.		M. longifolia Huds.	
Boswellia serrata Roxb.	Pelea madagascarica Baill.		M. spicata Huds.	
Myrtaceae	Dictamnus Fraxinella Pers.		Thymus citriodorus var. montanus	
Pimenta acris Wight.	Compositae		Compositae	
Labiatae	Artemisia caudata Michx.		Tagetes minuta L.	
Lophanthus anisatus Brenth.				
L. rugosa Fisch. & Mey.				
Collinsonia anisata				
Ocimum Basilicum L.				
O. sanctum L.				
O. gratissimum (L.) Boiss.				

Table 5—(Continued)

Methyl Chavicol	Anethole	Anisic Aldehyde	Carvone	Anisic Acid

Rutaceae
  Barosma serratifolia Willd.
  Dictamnus Fraxinella Pers.
  Barosma venustum
    Eckl. & Z.
  Clausena Anisum-olens
    (Bl.) Merr.
Rosaceae
  Prunus lusitanica Lois.
Compositae
  Solidago odora Ait.
  S. rugosa Mill.
  Artemisia caudata Michx.
  A. biennis Willd.
  A. Dracunculus L.
Piperaceae
  Piper Betel L.				

odor of rue due to methylnonylketone is masked by a pronounced odor of lemon due to the presence of citral. The larvae refused to eat these leaves. Likewise, they refused to eat filter paper scented with citral or limonene. Leaves of Rutaceae lacking methylnonyl-ketone were not sampled. It is possible that leaves containing it in quantities below threshold or leaves containing other constituents in excess likewise would be refused. The attracting quality of methylnonylketone is further illustrated by the readiness with which larvae sample filter paper soaked in the pure compound.

A few attempts were made to induce oviposition by adults of *P. ajax* by means of compounds attractive to the larvae. That these efforts did not meet with success might have been due to the absence of other requisite conditions, both external and internal. That the same attractants may serve feeding larvae and ovipositing adults is indicated by the ease with which the citrus-borer, *Citripestis sagittiferella*, can be induced to oviposit by the odor of citral, the same compound that serves as a feeding-type attractant of larvae.

It is clear that larvae of *P. ajax* are attracted by a group of chemicals representing several different types of compounds. Carvone is an aromatic ketone; coriandrol, an aliphatic terpene alcohol; methyl chavicol (para-oxyallylbenzene), an unsaturated phenol; anethole (para-methoxypropenylbenzene) also a phenol; anisic acid, para-methoxybenzoic acid (C_6H_4—$COCH_3$—$COOH$); anisic aldehyde, a methyl ether of para-oxybenzaldehyde; sedanolid, a lactone; and methylnonylketone, an aliphatic ketone.

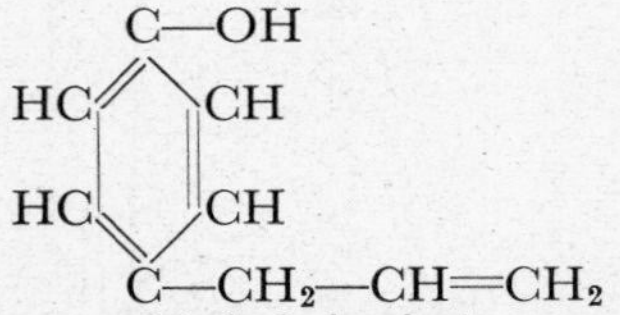

Methyl chavicol

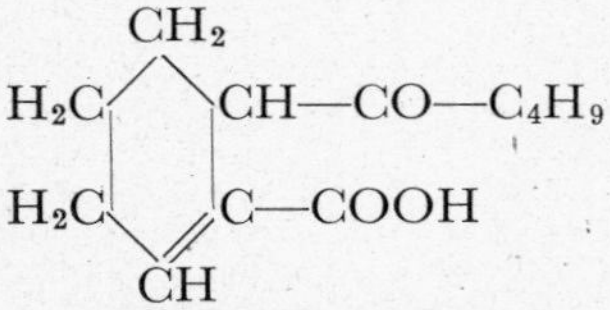

Sedanonic anhydride

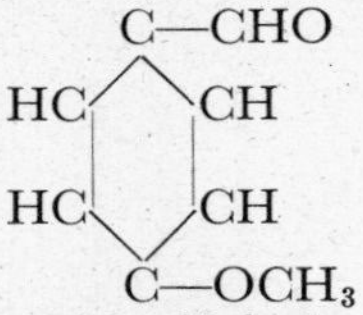

Anisic aldehyde

Organic Bases. Ordinarily essential oils do not possess organic bases in appreciable amounts. An outstanding exception is the cotton plant in which are present ammonia and trimethylamine in appreciable quantities. Also, there are emanations of these two compounds from the living plant (Power and Chesnut, 1925). Their great importance as attractants to cotton-feeding insects was early demonstrated by McIndoo (1926a) and by Folsom (1931). By means of the olfactometer described in Chapter 6, Folsom found that the Mexican boll weevil, *Anthonomus grandis* Boh., is attracted to the leaves, flowers, and bolls of the cotton plant, to cotton dew which contains by volume 74.42 parts of NH_4OH and 3.58 parts of trimethylamine in 1,000,000, and to NH_4OH and trimethylamine. Of the two chemicals, NH_4OH is the better attractant. In Tables 6, 7, and 8, the results of tests are expressed in terms of percentages

Table 6

RELATIVE ATTRACTIVENESS OF VARIOUS PARTS AND VARIETIES OF COTTON PLANTS TO THE MEXICAN BOLL WEEVIL*

Material	Positive	Negative	Number of Tests
Cotton plant leaves	58.6	41.3	6
Cotton plant flowers	60.8	39.0	17
Cotton plant squares	65.6	33.5	12
Cotton plant bolls	51.7	48.2	16
Cotton dew	76.9	23.0	5
Distilled water	64.7	35.2	89
Varieties of cotton:			
Wannamaker Big Boll	61.0	39.0	5
Mexican Tree Cotton	60.0	40.0	3
Sea Island Cotton	68.1	31.9	3
Pima Cotton	75.0	25.0	3

* After Folsom, 1931.

Table 7

RELATIVE ATTRACTIVENESS OF AMMONIA TO THE MEXICAN BOLL WEEVIL*

Experiment No.	Concentration	Positive	Negative	Number of Tests
101	1:1,000,000	77.0	22.9	1
102	500,000	58.3	41.6	1
103	200,000	54.1	45.8	1
104	100,000	45.8	54.1	1
105	10,000	31.2	68.7	1
106	1,000	18.7	81.2	1

* After Folsom, 1931.

of the total number of insects used in each test. Here again, as in preceding studies of naturally occurring attractants, it would ·be interesting to ascertain the effect of related compounds upon the insect. Would boll weevils be attracted, for example, to methylamine, dimethylamine, or even some of the diamines such as putrescine (tetramethylene diamine)? Also, how would the same weevils react toward leaves of *Chenopodium vulvaria* and the flowers of hawthorn, both of which give forth fishy ammoniacal odors by virtue of large concentrations of trimethylamine and propylamine?

Table 8

RELATIVE ATTRACTIVENESS OF TRIMETHYLAMINE TO THE MEXICAN BOLL
WEEVIL *

Experiment No.	Concentration	Positive	Negative	Number of Tests
116...........	1:1,000,000	64.5	35.4	1
117...........	500,000	56.2	43.7	1
118...........	200,000	58.3	41.6	1
119...........	100,000	43.7	56.2	1
120...........	10,000	31.2	68.7	1

* After Folsom, 1931.

Attractants in Peach. FOR CARPOPHAGOUS INSECTS. Analyses of attractants affecting fruit-feeding insects have produced a wealth of confusion because many such insects feed on fresh fruits in the larval stages only, the adult's menu leaning toward fermenting fruits and plant juices. The plum curculio, *Conotrachelus nenuphar* Hbst., is one carpophagous species which because of the similarity of its menu in all developmental stages is an ideal subject for study. Adults feed on leaves and young fruit; larvae, on fruit especially. Plum, peach, pear, apple, and cherry are the preferred plants. Fermenting baits are not attractive. Of these fruits, complete analyses of the odorous constituents have been made only of peaches (Power and Chesnut, 1921), but hardly enough experiments have been undertaken to determine which constituents serve as attractants. The odorous oil consists of linalyl esters of formic, acetic, valeric, and caprylic acids plus acetaldehyde and an aldehyde of unknown structure. Volatile acids may be present to some extent in the free state. Neither benzaldehyde nor HCN has been found in peach pulp; therefore, amygdalin is thought to be restricted to the kernel. The fraction of the oil to which is due the intense peachy odor is minute in quantity and unstable.

Snapp and Swingle (1929 and 1929a) studied the reactions of the plum curculio to steam distillates and fractional distillates of wild plum and cherry blossoms, peach fruit both green and ripe, and peach bark. In a McIndoo olfactometer these substances behaved as repellents. In field tests concentrated peach distillate is attractive. Quite commonly plant distillates act as repellents in an olfactometer when the opposite action is anticipated. Invariably this is due to maladjustment of concentration and inaptness of the apparatus for the particular job at hand.

To complete this study all of the odorous chemicals isolated from peach should be tested. Only the following have been tested in the field in evaporation cups: linalool, linalyl acetate, methyl alcohol, acetaldehyde, gallic acid, methyl formate, benzaldehyde, phenyl-hydrazone, phenylglycine, salicylaldehyde, and butyl acetate of which the last seven named exert some degree of attractiveness. Had it been possible at the time to investigate the unstable portion of the essential oil isolated from peach pulp, this might have been found to be the chief attrahent principle.

For Xylophagous Insects. Of borers which attack living trees or sound wood, many exhibit as decided preferences as phytophagous insects. Of the lamellicorn beetles, for example, *Cyllene pictus* feeds almost exclusively on hickory and *Callidium antennatum* prefers seasoned pine to spruce. Peach borers also are more or less restricted in their feeding habits, peach, and plum being favored species. Most attractants operating in the lives of borers are ovipositional in nature since larvae ensconced within their tunnels have little use for guidance; however, ovipositional-type attractants of adults are probably equally effective as feeding-type attractants of larvae. In either case, they occur in the form of odorous compounds in wood and bark. Very little information is available. Peach borers oviposit in the trunks of host trees whence emerging larvae bore into sapwood. Of many odorous substances tested, green peach distillate, ethyl benzoate, calcium malate plus acetaldehyde, sodium bisulfate, phenol, and salicylaldehyde are known to attract the lesser peach borer, *Synanthedon pictipes* G. & R., and the peach borer, *S. exitiosa* Say (Snapp and Swingle, 1929). Frost (1936) found that the best attractants one year were furfural (a heterocyclic compound) and piperonal (an aromatic ether with the odor of heliotrope) while the following year methyl cinnamate, amyl acetate, methyl alle-phenol, and sodium oleate excelled in that order.

OVIPOSITIONAL-TYPE ATTRACTANTS FOR ORIENTAL FRUIT AND CODLING MOTHS. With the exception of Siegler's (1940) work very little has been done with feeding-type attractants of carpophagous larvae. Efforts to induce hatching codling moth larvae to feed more readily on poisoned spray droplets by the addition of an attractant established the mild attractiveness of brown sugar, sorbitol, sucrose, corn syrup, d-fructose, glycerol, and malic acid. It seems quite certain, however, that these substances affect the sense of taste.

On the other hand certain of the results of experimental trapping of adults have unearthed some probable ovipositional-type attractants of the adults. The compounds listed in Table 9 are some of the

Table 9

RELATIVE ATTRACTIVENESS OF CERTAIN COMPOUNDS CHARACTERISTIC OF ESSENTIAL OILS TO THE CODLING MOTH AND THE ORIENTAL FRUIT MOTH

Codling Moth	Oriental Fruit Moth
1. Safrol	Linalool
2. Anethole	Safrol
3. Citronella oil	Citral
4. Clove oil	Anisic aldehyde
5. Fennel oil	Anisic acid
6. Limonene	Anethole
7. Citral	Furfural

more attractive ones not characteristic of fermenting baits (upon which the adults feed). They are the type commonly occurring in essential oils and until further evidence is forthcoming may reasonably be supposed to be ovipositional-type attractants. With the possible exception of furfural, which closely resembles benzaldehyde (common to Rosaceae) in odor, linalool (a common plant alcohol), citral, and limonene, these compounds have very little resemblance to those found in essential oils of peach, pear, plum, apple, and cherry. Correlation is lacking, and the problem is perplexing.

Aromatic Esters. The tobacco hornworm, *Protoparce sexta*, is not only restricted in its feeding habits, it shows a marked preference for certain plants chief among which is the jimson weed. Numerous attempts have been made to isolate the odorous principle of this plant (Morgan and Crumb, 1928; Morgan and Lyon, 1928; Gilmore and Milam, 1933). Although larvae of *P. sexta* feed readily on many species of plants after the leaves have been dipped into steam distillate of jimson weed leaves, not to mention sheets of paraffin impregnated with the same distillate, the compounds involved have not been identified. Random tests with more than 40 aromatic oils and ethereal salts turned up several powerful attractants. Of these amyl salicylate in an artificial plant attracted more adults than jimson weed itself. Other feeding-type attractants include benzyl benzoate, isoamyl benzoate, and isoamyl salicylate. Whether or not any of these compounds exist in jimson weed remains to be seen. Until now amyl salicylate has not been found free in nature.

Attractants in Corn. It is surprising that so little has been accomplished in searching for attractants in a plant as old, widespread, and economically important as corn. In 1922 McColloch demonstrated that the corn ear worm moth, *Heliothis obsoleta*, would oviposit on cotton twine soaked in fresh corn silk juice. In 1928 Moore found that the European corn borer, *Pyrausta nubilalis* Hbn., was attracted to steam distillates and petrol ether extracts of corn, smartweed, cocklebur, and greater ragweed. On the basis of leaf sections showing oil droplets between the epidermal cells he suggested that the odorous principle was an ethereal oil produced by certain of these cells. No further analyses have been reported.

Oxalic Acid. Oxalic acid is one of the more common so-called dibasic acids having the general formula $(CH_2)_n(COOH)_2$. Evidence confirming the importance of this acid and its salts as attractants of insects feeding on species of Polygonaceae has been highly suggestive but not yet conclusive. Verschaffelt's experiments with the beetle *Gastroidea viridula* Goeri were never completed. This insect in its adult and larval stages feeds on species of *Rumex*. In the Amsterdam Botanical Garden it attacked *R. scutatus* L., *Oxyria digyna* Hill, and cultivated *Rheum*. Leaves of *Lathyrus sylvestris*, an unacceptable plant which was never eaten fresh, were nibbled at slightly after having been immersed for some time in a normal solution of oxalic acid. In the absence of sufficient material it was impossible to determine whether the insect was attracted specifically by oxalic acid or by an acid medium in general. Its relations to other organic acids of the same series, malonic, succinic, etc., as well as to the hydroxy acids common to unripe fruit, malic, citric, etc., should be tested.

Larvae of the American Copper butterfly, *Lycaena hypophlaeas* Bdv., are without a doubt the ideal experimental insects for solving the rôle of oxalic acid in food-plant selection since they are known to eat but a single plant, *Rumex Acetosella*. It is conceivable that a very fine adjustment may exist between these insects and the concentration or amount of oxalic acid and that different species of *Rumex* may contain the acid in different concentrations.

Solanaceae. Some of the most thorough experiments to date have been those of Raucourt and Trouvelot (1936) which dealt with the constituents of potato plant leaves attractive to the Colorado potato beetle larva. While these workers did not succeed in identifying the attractive principle, they determined many of its properties and reactions. It belongs to the nonliquid portion of the cell contents and is expressed from the leaves simultaneously with the chlorophyll. It is soluble in hot alcohol. After removal of fatty matter from the alcohol extract it becomes soluble in water. It is nonvolatile and very stable, surviving several days at 100° C. There is reason to believe that it is a nitrogenous principle but neither a saponifiable lipid nor a solanin.

A rather nice relationship was found between the attractiveness of this principle and its concentration. Plant consumption by larvae was found by actual test to be proportional to concentration and negligible when this was one-fourth normal, illustrating again the importance of concentration. It was found also that natural fermentation produced repellents in leaf juices which destroyed the attractive principle and prevented consumption of the leaves by the larvae.

Risler's (1941) observation that a 5 per cent solution of $(NH_4)_2CO_3$ attracts *Leptinotarsa decimlineata* is of considerable interest, but throws no additional light on the nature of the attractive principle of potato leaves.

Lactones. Certain other insect-plant-chemical relations have been the subject of considerable study but as yet are not completely understood. Two of these are of interest because of their resemblance on the one hand and their contrast on the other. The Monarch butterfly, *Danais plexippus* L., and the Painted Lady, *Nymphalis* *cardui*, are practically cosmopolitan. Larvae of the former are greatly restricted in feeding habits, preferring species of *Asclepias* in some parts of their range and other Asclepiadaceae and Apocynaceae in other parts, while the latter feeds on a wide variety of plants. Attempts to ascertain and test the attractive principle of milkweed have not been highly successful. Matsurevich (1936), Rheineck (1939), and others have analyzed species of *Asclepias* but determined the characteristic odor merely as probably due to a lactone or a special phytosterol. Whatever the compound, it is notably present in the latex which is most attractive to larvae.

Of the plants attacked by larvae of *Nymphalis cardui* one, Everlasting (*Gnaphalium obtusifolium* L.), is especially interesting because of its peculiar and penetrating odor. In his analysis of the plant Jannke (1938) isolated "a soft yellow substance whose delicate nut-like odor was identical with the plant itself." Presumably this is a phenol, enolic acid, or an enolic lactone. Preliminary tests of the substance with larvae of *N. antiopa* indicated that it is probably the principle which identifies the plant to larvae (Dethier, unpublished). More detailed experimentation has been limited by the minute amount of material available for study.

Geraniol. The saga of the attractant geraniol and the Japanese beetle scarcely bears repetition (cf. p. 176). Geraniol, for which the beetle has so great an affinity, is a cyclic alcohol closely related to cyclic terpenes and is widely distributed throughout the plant kingdom. Why this particular compound should be so attractive and so nearly specific for the Japanese beetle, a polyphagous species, is an unsolved riddle. There are many other polyphagous species of similar habits that it fails to attract. On the other hand it is attractive— as are also *Acacia* flower oil, linalool, and geranyl acetate—to *Acacia*-feeding larvae of the bagworm, *Acanthopsyche junodi* Heyl. (Ripley, Petty, and van Heerden, 1939).

Discussion. Certain facts yielded by these studies stand out clearly. They permit of a few generalizations regarding leaf-feeding insects which may extend wholly or partially to other feeding categories. It is now well established that plants are recognized by the feeding stages of insects through the agency of odors. These emanate from an extremely complex mixture of compounds which, in the case of healthy living plants, are comprehended under the term essential oils. While a blending of materials is indeed responsible for the odor of a plant, usually some one chemical imparts the identifying characteristic quality. Frequently this compound may be isolated. The experiments discussed illustrate the method of approach and the problems encountered in this type of work. Having once found the compound essentially responsible for an odor, no mean undertaking in itself, it remains to test the reactions of the insect in question to this compound. Here also many difficulties are encountered. The frequent failure of plant distillates to act as anticipated arises, among other things, from a faulty adjustment of concentration plus the fact, already noted, that natural plant odors arise from exquisitely blended mixtures. Manhandling by the research worker upsets this arrangement. Commonly the odor of a steam distillate in no way resembles that of the living plant. Since, as indicated in the afore-cited experiments, odor alone is the guiding character, failure of an insect to respond to a distillate as expected is no indication that this does not contain attractive compounds.

One should be impressed with the bewildering number of compounds available in essential oils as odor signposts to leaf-feeding insects and by their complexity. Nor is there any conformity or restriction to the type of compound that may act as an attractant. While oxygenated constituents predominate, especially ketones and aldehydes, the number and kinds of other compounds from which to choose are legion.

REFERENCES

Atkinson, D. J.: Some experiments on the control of the bamboo shot-hole borer, *Dinoderus* spp. in bamboo dunnage, *Burma Forest Bull.*, **32**, 1–14 (1936).

Belzung, E.: "Anatomie et Physiologie Végétales," Paris, Germer Baillière & Co., 1900.

Carl, A. L., and E. K. Nelson: The occurrence of citric and isocitric acid in blackberries and in dewberry hybrids, *J. Agr. Research*, **67**(7), 301–303 (1943).

Czapek, F.: "Biochemie der Pflanzen," Jena, Gustav Fischer, 1921, v. 3.

Delange, R.: "Essences naturelles et parfums," Paris, A. Colin, 1930.

Dethier, V. G.: Gustation and olfaction in lepidopterous larvae, *Biol. Bull.*, **72**(1), 7–23 (1937).

——: Chemical factors determining the choice of food plants by *Papilio* larvae, *Am. Naturalist*, **75**, 61–73 (1941).

Dumont, C.: Expériences sur la modification profonde du régime alimentaire de diverses chenilles, *Ann. soc. entomol. France*, **97**, 59–104 (1928).

Finnemore, H.: "The Essential Oils," New York, D. van Nostrand Co., 1926.

Folsom, J. W.: A chemotropometer, *J. Econ. Entomol.*, **24**(4), 827–833 (1931).

Frost, S. W.: Tests on baits for oriental fruit moth, 1935, *Ibid.*, **29**(4), 757–760 (1936).

Gildemeister, E., and Fr. Hoffmann: "The Volatile Oils," Milwaukee, Pharmaceutical Review Pub. Co., 1900.

Gilmore, J. U., and J. Milam: Tartar emetic as a poison for the tobacco hornworm moths, a preliminary report, *J. Econ. Entomol.*, **26**(1), 227–233 (1933).

Gortner, R. A.: "Outlines of Biochemistry," 2d ed., New York, John Wiley & Sons, 1938.

Greshoff, M.: Sur la distribution de l'acide cyanhydrique dans le règne végétal, *Bull. sci. pharmacol.*, **13**(11), 589–602 (1906).

Haberlandt, G.: "Physiologische Pflanzenanatomie," Leipzig, W. Engelmann, 1904.

Hampton, F. A.: "The Scent of Flowers and Leaves," London, Dulau & Co. Ltd., 1925.

Hansberry, R.: Testing stomach insecticides in "Laboratory Procedures in Studies of the Chemical Control of Insects," American Association for the Advancement of Science, Washington, D. C., 1943, pp. 85–94.

Hogg, P. G., and H. L. Ahlgren: Environmental, breeding, and inheritance studies of hydrocyanic acid in *Sorghum vulgare* var. *sudanense, J. Agr. Research*, **67**(5), 195–210 (1943).

Hooker, W. J., J. C. Walker, and K. P. Link: Effects of two mustard oils on *Plasmodiophora brassicae* and their relation to resistance to clubroot, *Ibid.*, **70**(3), 63–78 (1945).

Hoskins, W. M., and R. Craig: The olfactory responses of flies in a new type of insect olfactometer, *J. Econ. Entomol.*, **27**(5), 1029–1036 (1934).

Jannke, P. J.: Phytochemical analysis of *Gnaphalium obtusifolium* Linné, *Pharm. Arch.*, **9**(2–5), 17–75 (1938).

Langford, G. S., M. H. Muma, and E. N. Cory: Attractiveness of certain plant constituents to the Japanese beetle, *J. Econ. Entomol.*, **36**(2), 248–252 (1943).

Leblois, A.: Recherches sur l'origine et le développement des canaux sécréteurs et des poches sécrétrices, *Ann. sci. nat. Botan.*, ser. 7, **6**, 247–330 (1887).

McColloch, J. W.: The attraction of *Chloridea obsoleta* Fabr. to the corn plant, *Ibid.*, **15**(5), 333–339 (1922).

McIndoo, N. E.: An insect olfactometer, Ibid., **19**(3), 545–571 (1926).

——: Senses of the cotton boll weevil. An attempt to explain how plants attract insects by smell, *J. Agr. Research*, **33**(12), 1095–1141 (1926a).

——: The relative attractiveness of certain solanaceous plants to the Colorado potato beetle, *Leptinotarsa decemlineata* Say, *Proc. Entomol. Soc. Wash.*, **37**(2), 248–252 (1935).

* Matsurevich, I. K.: *Univ. état Kiev, Bull. sci., Receuil chim.*, **2**(2), 9–19 (1936).

Moore, R. H.: Odorous constituents of the corn plant in their relation to the European corn borer, *Univ. Oklahoma Bull.*, **8**(30), 16–18 (1928).

Morgan, A. C., and S. E. Crumb: Notes on the chemotropic responses of certain insects, *J. Econ. Entomol.*, **21**(6), 913–920 (1928).

——, and S. C. Lyon: Notes on amyl salicylate as an attrahent to the tobacco hornworm moth, *Ibid.*, **21**(1), 189–191 (1928).

Mosher, F. H.: Food plants of the gipsy moth in America, *U.S. Dep. Agr. Bull.* 250, 1–39 (1915).

Parry, E. J.: "The Chemistry of Essential Oils and Artificial Perfumes," 3d ed., London, Scott, Greenwood and Son., 1918, v. 1, 2.

Peterson, A.: Some chemicals attractive to adults of the onion maggot (*Hylemyia antiqua* Meig.) and the seed corn maggot (*Hylemyia cilicrura* Rond.), *J. Econ. Entomol.*, **17**(1), 87–94 (1924).

Power, F. B.: The distribution and characters of some of the odorous principles of plants, *Ind. Eng. Chem.*, **11**(4), 344–352 (1919).

——, and V. K. Chesnut: The odorous constituents of peaches, *J. Am. Chem. Soc.*, **43**, 1725–1738 (1921).

——, and ——: The odorous constituents of the cotton plant. Emanation of ammonia and trimethylamine from the living plant, *Ibid.*, **47**, 1751–1774 (1925).

Ramsay, J. A., C. G. Butler, and J. H. Sang: The humidity gradient at the surface of a transpiring leaf, *J. Exp. Biol.*, **15**(2), 255–265 (1938).

Raucourt, M., and B. Trouvelot: 1936. Les principes constituants de la pomme de terre et le doryphore, *Ann. épiphyt. phytogéné.*, **2**(1), 51–98 (1936).

Rheineck, A. E.: A phytochemical study of *Asclepias syriaca* Linné, *Pharmaceut. Arch.*, **10**(4), 53–64; **10**(5), 69–80; **10**(6), 93–98 (1939).

Rhin, A. E., and H. G. Dekay: The glycosides of *Asclepias cornuti* or the common milkweed, *J. Am. Pharm. Assoc.*, **29**(2), 69–71 (1940).

Ripley, L. B., B. K. Petty, and P. W. van Heerden: Further studies on gustatory reactions of the wattle bagworm (*Acanthopsyche junodi* Hey), *Union S. Africa, Dep. Agr., Forestry Sci. Bull.*, 205, 1–20 (1939).

Risler, J.: La lutte contre le doryphore. Recherches expérimentales. Applications pratiques, *Compt. rend. acad. Agr. France*, **27**(20), 1110–1120 (1941).

Schwarz, E.: The reason why Catocola eggs are occasionally deposited on plants upon which the larvae cannot survive; and a new variation (Lepid., Noctuidae), Entomol. News, **34**(9), 272–273 (1923).

Siegler, E. H.: Laboratory studies of codling moth larval attractants, *J. Econ. Entomol.*, **33**(2), 342–345 (1940).

Snapp, O. I., and H. S. Swingle: Preliminary report on attrahents for peach insects, *Ibid.*, **22**(1), 98–101 (1929).

——, and ——: Further results with the McIndoo Insect Olfactometer, *Ibid.*, **22**(6), 984 (1929a).

Stahmann, M. A., K. P. Link, and J. C. Walker: Mustard oils in crucifers and their relation to resistance to clubroot, *J. Agr. Research*, **67**(2), 49–63 (1943).

Tschirch, A.: "Angewandte Pflanzenanatomie," Vienna and Leipzig, Urban & Schwarzenberg, 1889, 2 vols.

——, and O. Oesterle: "Anatomischer Atlas der Pharmakognosie und Nahrungsmittelkunde," Leipzig, C. H. Tauchnitz, 1900.

Van Someren, V. G. L.: Chemical changes in the food-plant as a cause of failure in rearing larvae, *Proc. Roy. Entomol. Soc. London*, A, **12**(1–2), 10 (1937).

Verschaffelt, E.: The cause determining the selection of food in some herbivorous insects, *Proc. Acad. Sci. Amsterdam*, **13**(1), 536–542 (1910).

Wehmer, C.: "Die Pflanzenstoffe," Jena, G. Fischer, 1929–1935, vols. 1, 2, and supplement.

Chapter 4

Fermentation Products (Alcohols, Acids, Aldehydes, Esters, Carbinols)

Source of Attractants. Of the many organisms dependent upon higher plants, plant products, animals, and animal products for subsistence, none produce such outstanding chemical changes in their nutrient media as do bacteria and fungi. The action of these on the numerous substrates provided by other organisms results in

Table 10

NATURALLY OCCURRING DISACCHARIDES*

Name	Source
Glucoxylose	Leaves and branches of *Daviesia latifolia*
Arabobiose	Arabinic acids
Strophanthobiose	In the strophanthus glucosides
Rutinose	In the glucosides rutin and datisein
Primverose	In the glucosides primocrin, gentiaraulin, rhamnicoside
Vicianose	In the glucosides gein, vicianin, etc.
Trehalose	In fungi
Maltose	Degradation of starch
Gentiobiose	In the glucoside amygdalin and the trisaccharide gentianose
Dextrinose	In beer, urine, blood, honey, liver, etc.
Cellobiose	Degradation of cellulose
Isocellobiose	Degradation of cellulose
Sucrose	Common in plants
Turanose	In the trisaccharide, melezitose
Lactose	In the milk of all mammals
Melibiose	In the trisaccharide raffinose

*After Harrow and Sherwin: "A Textbook of Biochemistry," Philadelphia, W. B. Saunders Co., 1935.

the production of many compounds suitable either as food for insects or as attractants indicating the presence of food or oviposition sites.

With reference to these substrates it is frequently expedient to distinguish between the nonnitrogenous and the nitrogenous. In nature the most readily available nonnitrogenous substrates are vege-

75

table in nature while nitrogenous substrates are more commonly of animal origin. The term fermentation cannot be correctly reserved for those reactions taking place on nitrogen-free substrates alone since amino acids also may be fermented and, too, the amount of alcohol formed during fermentation is dependent upon the amino acid content of the media. In presenting this chapter as titled, cognizance of these facts has been taken; therefore, the present chapter treats not only of attractants arising from the fermentation of plants and plant products but also those from the fermentation of proteins contributed by the participating bacteria and fungi as well. The next chapter will consider attractants arising from the decomposition of animals and animal products, a few outstanding cases of fermentation of animal substrate, and certain decomposition products of some fungi and algae.

Table 11

NATURALLY OCCURRING TRISACCHARIDES*

Name	*Source*
Rhamninose........	In the glucosides sophorin and xanthorhamnin
Robinose...........	In the glucoside robinin
Manninotriose......	Ash-tree manna
Levidulinose........	In Konjak, mannan
Raffinose...........	In sugar beet and cotton seed
Gentianose.........	In gentian roots
Melezitose..........	In the manna of Douglas firs, honeydew

* After Harrow and Sherwin: "A Textbook of Biochemistry," Philadelphia, W. B. Saunders Co., 1935.

The classical conception of fermentation as involving the production of alcohol by the anaerobic metabolism of yeasts was expressed by the equation:

$$C_6H_{12}O_6 \rightarrow 2CO_2 + 2C_2H_5OH$$

Glucose   **Ethyl alcohol**

Since Pasteur's demonstration of the rôle of yeasts in fermentation many of the lower plants have been shown to be the agents of fermentation of innumerable substrates. Of fundamental plant products the most important substrates by and large are the nonnitrog-

Table 12

SOME OF THE MORE IMPORTANT PRODUCTS OF FERMENTATION BY BACTERIA
AND FUNGI

Organism	Substrate	Products
BACTERIA		
Bacterium xylinum	Aldoses	Monobasic acids
	Alcohols	Ketoses
B. acidi lactici	Glucose	Lactic acid
B. coli	"	"
B. coli communis	"	"
B. lactis aerogenes	"	"
B. cloacae	"	"
Lactobacillus pentoaceticus	"	"
Bacterium tartaricus	Peptone with carbohydrate	l-Acetylmethylcarbinol
B. subtilis	"	"
Tyrothrix tenius	"	"
Bacterium asiaticus mobilis	. .	Butyleneglycol
B. cellulose dissolvens	Cellulose	Acetic acid, butyric acid, ethyl alcohol, H_2, and CO_2
Several organisms	Cellulose	Methane and hydrogen
	Sugar	Formic acid breaking down into H_2 and CO_2
All mesophilic anaerobes	Pectins	Acetic acid
(e.g., B. amylobacter)	Starches	Acetone
	Hexoses	Butyl alcohol
	Pentoses, etc.	
Granulobacter pectinovorum	Hexose	Ethyl alcohol, butyl alcohol, acetone, acetic acid, formic acid, H_2, and CO_2
and others		
Butyric acid bacteria	Glucose	Butyric acid, acetic acid, H_2, and CO_2
Many bacteria, especially those	Carbohydrates	Propionic acid
occurring in milk and cheese	Lactic acid	Acetic acid
	Tartaric acid	Carbonic acid
	Quinic acid	
	Glycerol	
	Proteins and their degradation products	
	Glycerol	n-Butyl alcohol
	Mannitol	n-Butyric acid
	Starch	Ethyl alcohol
	Invert sugar	Caproic acid ?
	Lactate	Propionic acid
		Butyric acid
Bacterium mesentericus	. .	Isovaleric and oleic acid
Acetobacter xylinum	Glycerol	Dihydroxyacetone
A. melanogenum	"	"
A. gluconicum	"	"
A. xylinoides	"	"
A. orleanse	"	"
A. aceti	"	"

Table 12—(Continued)

Organism	Substrate	Products
FUNGI		
Yeasts	Leucine plus isoleucine plus sugar	Amyl alcohols
Yeasts (by condensation)	Acetaldehyde Benzaldehyde	1-Phenylacetylcarbinol
Yeasts (at expense of own protein)	"	Isoamyl alcohol, *d*-amyl alcohol, propyl alcohol, isobutyl alcohol
Aspergillus elegans	Glucose	Mannitol
A. nidulans	Mannose	"
A. Wentii	Galactose	"
P. chrysogenum	Xylose	"
Helminthosporium geniculatum	Arabinose	"
Clasterosporium	"	"
Mucor racemosus	Sucrose	Glycerol
A. niger	"	"
Aspergillus sp.	Glucose	Glycerol
Helminthosporium geniculatum	"	"
Clasterosporium	"	"
A. Wentii	"	"
P. griseo-fulvum	"	6-hydroxy-2-methyl-benzoic acid and 2-5-dihydroxybenzoic acid
Fusarium lini	Any sugar	Ethyl alcohol and H_2O
Aspergillus niger	A variety	Oxalic acid
A. violaceofuscus	"	"
A. ochraceus	"	"
Penicillium oxalicum	"	"
Many spp. and genera (e.g., *P. luteum*)	Many	Citric acid
Species of *Penicillium* and *Aspergillus*	Several	*d*-gluconic acid
P. purpurogenum	Media containing *d*-mannose	*d*-mannonic acid
Rhizopus nigricans	. .	Fumaric acid
A. fumaricus	Carbohydrates	Malic acid
Clasterosporium	"	"
Mucor mucedo	"	Succinic acid
P. spiculisporum and others	"	"
Mucor	Sugars	Ethyl alcohol
Rhizopus	"	"
Aspergillus	"	"
Penicillium	"	"
Monilia	Glucose	Acetaldehyde
Mucor	Sucrose	"
Aspergillus	"	"
Penicillium	"	"
Oidium	"	"
P. digitatum	Glucose	Ethyl acetate

enous ones, especially carbohydrates. These may be classified as follows:

1. Monosaccharides
   Aldose sugars
   Ketose sugars
   Monoses to decoses
2. Compound sugars
   Disaccharides
   Trisaccharides
   Tetrasaccharides
   Polysaccharides (nonsugars)

Of these, pentoses and hexoses are the most important. The former occur most commonly as polysaccharides. Organisms capable of fermenting them are common in nature.

Table 13

OCCURRENCE OF SOME SUGAR ALCOHOLS
IN NATURE

Sugar Alcohols	Occurrence
Glycerol	In fats
Erythritol	As esters in some algae
Mannitol	Sap, leaves, fruit, etc.
Sorbitol	Fruits of most Rosaceae
Dulcitol	In certain plants
Iditol	In certain plants

The organisms whose enzymes perform fermentation on these substrates are numbered among the lower plants and include bacteria, yeasts, and molds. Some species are rigidly specific in action. As a matter of fact, certain bacteria are specific for some rare sugars to such an extent that the interaction of the two is frequently employed as a means of identification of one or the other. Other plants ferment any number of sugars with equal facility. Molds, under optimum conditions, will eventually oxidize carbohydrates and the products of their own metabolism to carbon dioxide and water. The intermediate compounds, each of which may serve as a substrate for some other species, include acids of the fixed type (gluconic, citric, fumaric, oxalic, etc.), fatty acids, aromatic acids, lactonic acids, phenolic acids, simple alcohols, carbinols, glycols, aldehydes, ketones, pyrone, oxyanthraquinones, quinone derivatives, more complex carbohydrates, fats, and sterols. Bacteria are more prone to produce volatile acids (acetic, propionic, butyric, lactic, etc.). In addition to these compounds others may arise from the action of fermentation products on sugars already present. Aldoses and ketoses can condense to form esters (with acids), ethers (with alcohols), and acetals (with aldehydes or ketones), all of which may in turn serve as substrates. Sugar alcohols, of wide occurrence in plants (Table 13), may be further changed, not fermented, by bacteria and fungi.

Types of Insects Responding to These Attractants. These then, are a few of the compounds that must be considered as possible attractants. The insects most likely to encounter them either as feeding-type attractants or as ovipositional-type attractants are carpophagous and xylophagous species, honey-feeders, some sap-feeders, and some leaf-feeders. Lepidoptera, Coleoptera, and Diptera predominate. Of Lepidoptera, carpophagous species are chiefly representatives of the Eucosmidae. The most important xylophagous species are the clear-winged moths (Aegeriidae). Most carpophagous Coleoptera are weevils (Curculioninae) which feed principally on fruits, seeds, and nuts. Xylophagous species are represented chiefly by the families Scolytidae, Cerambycidae, and Buprestidae. Carpophagous Diptera belong to the families Trypetidae and Drosophilidae. Altogether five species of fruit and orchard insects, exclusive of the fruit flies, have been the objects of considerable study. They are the codling moth (*Carpocapsa pomonella* L.), the oriental fruit moth (*Grapholitha molesta* Busck), the peach borer (*Synanthedon exitiosa* Say), the lesser peach borer (*S. pictipes* G. and R.), and the plum curculio (*Conotrachelus nenuphar* Hbst.). Codling moths oviposit on apples near the calyx or on adjacent foliage. Hatching larvae bore into the fruit by way of the calyx. Peach borers oviposit on the trunks of trees whence newly hatched larvae bore into sound wood. Moths of the groups mentioned feed upon nectar, exuding sap, fermenting fruit, etc. Eggs of the plum curculio are deposited usually on stone fruits which serve as food for the larvae. Adults feed on foliage and young fruit of plum, pear, apple, and cherry.

Obviously there are more than one type of attractant operating here; namely, ovipositional-type attractants of adults, feeding-type attractants of adults, and feeding-type attractants of larvae. The chemical aspect of ovipositional-type attractants involves compounds present in foliage and unripened fruit or in the case of peach borers, in foliage and bark. Feeding-type attractants of the larvae— also adults in the case of the plum curculio—are probably identical. In any case they are essential oils. The work of Snapp and Swingle (see p. 67) represents one of the few careful studies on this subject. The feeding-type attractants of moths are of an entirely different nature and are truly fermentation products.

Codling Moth Attractants. Armed with the traditional knowledge that baits composed of fermenting carbohydrates, sugars, molasses, beer, etc., attracted flying insects, especially moths, and that various essences should attract insects, investigators were soon trying every possible combination of substances in a search for a bait that would attract all comers. Hundreds of aromatic compounds, chosen more or less at random, were tested. Since most of these attempts were prompted by economic motives, little was done to systematize the work. The first serious attempt to evaluate sweetened baits was that of Peterson (1925). It was apparent from his experiments that molasses-yeast baits owed their attractiveness to fermentation products, especially since they became more attractive after a period of one to four days. This was the status of the problem when in 1931 Eyer and his associates resumed where Peterson left off and exposed a vast new panorama of important phenomena. In early studies they (Eyer, 1931; Eyer and Rhodes, 1931; Eyer, 1934) began analyzing molasses baits which were known to be extremely attractive to *Carpocapsa pomonella* L. Several important preliminary facts came to light. (1) The attractive principle is most closely associated with sugar decomposition prior to alcohol and acetic acid formation, a fact which suggests esters. (2) The formation of alcohol and gas is of secondary importance. (3) Conversion to acetic acid by yeast renders the mixture repellent; thus the addition of yeast decreases the effectiveness of bait while the addition of a preservative such as sodium benzoate prolongs its effectiveness. (4) Attractiveness apparently is associated with composition, boiling point, and odor. To check the first suggestion, 35 esters were tested. The following showed promise in this order of attractiveness: isobutyl phenyl acetate, geranyl formate, ethyl oxyhydrate, diphenyl oxide, citronellal, bromostyrol, and methyl cinnamate. Esters of malic acid were but slightly attractive and commercially prepared extracts of apple blossoms or fruit were ineffective.

While the agency of bacteria in the production of these and similar compounds was known, its extreme importance was not fully appreciated until Eyer, Medler, and Linton (1937) inoculated pure carbohydrate substrates with pure cultures of certain species to produce a bait superior in attractiveness to those allowed to ferment in the field through natural contamination. Of approximately 265 species of bacteria and yeasts comprising the normal flora of baits, three because of their constant occurrence throughout the season were subjected to intensive study. The three were *Aerobacter aero-*

Table 14

SUMMARY OF TESTS OF HETEROFERMENTATIVE BACTERIA IN CODLING MOTH BAITS*

Series I—Pure cultures of organisms in 250 cc. of carbohydrate solution placed in gauze-covered pint jars surrounded by gallon jars of water

Name of Organism and Carbohydrate	Total Moths Caught	Number of Jars Used	Duration of Experiment (Days)	Average Number of Moths per Jar per Day	Per Cent of Total Catch	Percentage Ratio to Cane-sirup Check
Aerobacillus polymyxa in 10% sucrose in water with 1% peptone........	9	2	35	0.13	3.75	40.91
Same in 10% peptonized dextrose...	31	2	35	0.44	12.92	140.91
Same in 10% "Brer Rabbit" sirup..	40	4	35	0.29	16.67	181.82
Aerobacter aerogenes in 10% sucrose in water with 1% peptone.........	39	3	35	0.37	16.25	177.27
Same in peptonized dextrose.......	10	3	35	0.10	4.17	45.45
Same in cane sirup..............	89	4	35	0.64	37.08	404.55
Cane-sirup check................	22	2	35	0.31	9.17	100.00

Series II—As in Series I except mixed cultures of the two organisms placed in 500 cc. of carbohydrate solution in uncovered quart jars

	Total Moths Caught	Number of Jars Used	Duration of Experiment (Days)	Average Number of Moths per Jar per Day	Per Cent of Total Catch	Percentage Ratio to Cane-sirup Check
A. polymyxa and *aerogenes* in 10% peptonized sucrose................	27	5	35	0.15	5.13	13.17
Same in 10% peptonized dextrose..	50	5	35	0.29	9.15	24.39
Same in 10% cane sirup..........	244	5	35	1.39	46.39	119.02
Cane-sirup check................	205	5	35	1.17	38.97	100.00

* After Garcia, 1941.

genes, *A. oxytocum*, and *Aerobacillus polymyxa*. The superiority of baits fermented by pure cultures of these species, termed heterofermentative because of the wide variety of compounds they produce, is indicated by the results tabulated in Table 14 (Garcia, 1941). Protein deficiency necessitated the addition of peptone to promote proper growth in cultures.

On substrates of glucose, sucrose, or fructose species of *Aerobacter* produce acetaldehyde, formic, acetic, succinic, and lactic acids, ethyl alcohol, acetyl methyl carbinol, carbon dioxide, and hydrogen. *Aerobacillus* gives rise to the same fermentation products plus butyl alcohol, butylene glycol, and acetone. Secondary reactions may yield fumaric and lactic acids which may in turn be converted to malic acid. Still other reactions involving proteins contributed by the organisms themselves may yield amino and phenyl substitution compounds of which α-amino acetic acid (glycine), phenylacetic acid, and hydroxybenzoic acids are examples.

Investigations of baits fermented by pure cultures of bacteria have thus far followed two courses: (1) determination of those compounds present in appreciable amounts during the period of maximum attractiveness of baits to the codling moth, namely, within the initial 10 days of exposure in the field, and (2) testing of the attractiveness of related chemicals to the codling moth. In the first instance the procedure consisted of inoculating 500 cc. samples of sterile cane-

Table 15

ANALYSES OF CANE-SIRUP BAITS FERMENTED BY PURE CULTURES OF AEROBACTER
AND AEROBACILLUS FOR 10 DAYS AT ROOM TEMPERATURE *

Bacterium	Ethyl Alcohol	Iso-propyl Alcohol	Butyl Alcohol	Acetic Acid	Pro-pionic Acid	Butyric Acid	Lactic Acid	Acetone
Aerobacter aerogenes								
Strain 1	0.82%	..	..	0.69%	..	..	..	..
Strain 2	0.72	..	..	0.61	..	..	..	Trace
Strain 3	1.12	..	..	0.51	..	..	..	Trace
Strain 4	1.08	..	..	0.59	..	..	..	Trace
A. oxytocum	1.04	..	..	0.70	..	..	..	Trace
Aerobacillus polymyxa								
Strain 1	0.73	..	..	0.60	..	..	0.12	0.038
Strain 2	0.82	..	..	0.59	..	..	0.10	0.036

* After Eyer and Medler, 1940.

sirup solution with each of the three species of bacteria mentioned above. The samples were then analyzed at intervals while being allowed to ferment at room temperature for 10 days. Results as listed in Table 15 show that ethyl alcohol is the most abundant simple alcohol and acetic and lactic acids the chief acids. Furthermore, it was found that aldol and acyloin condensations yielded measurable amounts of acetyl methyl carbinol and acetone.

Table 16

RESULTS OF COMPARISONS OF SIMPLE ALCOHOLS AND ACIDS
IN AN OLFACTOMETER*

Material	Number of Moths Caught	Average Catch per Replication	Percentage Ratio to Check
ALCOHOLS			
Methyl.........	1	0.33	100
Ethyl..........	0	..	..
n-Propyl........	2	0.66	200
Isopropyl......	4	1.33	400
n-Butyl.........	0	..	..
Isobutyl........	0	..	..
Amyl...........	0	..	..
Isoamyl........	0	..	..
Check.........	1	0.33	100
ACIDS			
Formic.........	1	0.33	50
Acetic.........	0	..	..
Propionic.......	2	0.66	100
Butyric........	0	..	..
Valeric........	(Not tested)		
Check.........	2	0.66	100

* After Eyer and Medler, 1940.

Testing the relative attractiveness of related chemicals was the logical sequence to this work. The following homologous series were tested: (1) alcohols related to ethyl alcohol, (2) organic acids related to acetic acid, (3) esters capable of being formed by interaction of the aforementioned alcohols and acids, (4) dibasic and hydroxy acids related to lactic acid and their more common alkyl ester derivatives, (5) higher alcohols related to acetyl methyl carbinol, and (6) gases formed during fermentation.

Compounds to be compared were tested by several procedures. Each compound was emulsified at the rate of 1 cc., or 1 g. in the case of solids, to 99 cc. of a mixture of gum tragacanth (200 g.), glycerin (100 cc.), and water (to make 2 liters). First, emulsions were placed in jars fitted to an olfactometer equipped with eight lateral arms (Eyer, 1937). The attractive value of those showing promise was then compared by exposing 1 per cent emulsions in jars suspended in an apple orchard. Cane-sirup bait served as a control. Procedure two was modified also in that pails were hung in an apple tree completely enclosed by screen. Newly emerged moths were used as experimental animals.

Table 17

RESULTS OF COMPARISONS OF HYDROXY, DIBASIC, AND AROMATIC ACIDS
IN AN OLFACTOMETER *

Material	Number of Moths Caught	Average Catch per Replication	Percentage Ratio to Check
HYDROXY AND DIBASIC ACIDS			
Oxalic....................	0	..	..
Lactic....................	1	0.33	66
Malonic...................	1	0.33	66
Succinic..................	2	1.00	200
Succinic (solid)...........	1	0.50	100
Check....................	1	0.50	100
AROMATIC ACIDS			
Benzoic..................	0	..	..
Benzoic (solid)...........	0	..	..
α-Toluic.................	0	..	..
α-Toluic (solid)..........	2	200	200
Check...................	1	100	100

* After Eyer and Medler, 1940.

In nature codling moths are crepuscular, being most active during early evening at temperatures slightly above 70° F. In order to duplicate natural conditions as closely as possible in the olfactometer, tests were conducted in a dark room at temperatures of 70° to 75° F. from the hours of 0630 to 2200. Indirect lighting of low intensity (1 foot candle) illuminated the eight lateral arms of the apparatus while a small fan sucked air down through the floor of the central chamber to maintain circulation. The results listed in Tables 16, 17, 18, and 19 were obtained with this set-up and are expressed in terms of percentage ratio to check. Ratios between 50 and 100 are considered just slightly attractive while in field tests

(Tables 20, 21, 22, 23, 24) any value above 0 is considered attractive inasmuch as the control, cane sirup, is itself very attractive. Of the alcohols, n-propyl and isopropyl are the most attractive; of the simple acids, propionic (Table 16); of the hydroxy and dibasic acids, succinic; of the aromatic acids, α-toluic (Table 17); of the esters, the methyl series (when the alcohol radicals are compared) and the esters of propionic acid (when the acid radicals are compared). Butyl esters and esters of butyric acid are also considerably attractive. In considering the attractiveness of the methyl esters it is interesting to observe that there is a progression in attractiveness as follows: esters of lactic acid (133) < malonic (160) < benzoic (160)

Table 18

RESULTS OF COMPARISONS OF ESTERS OF SIMPLE ACIDS IN AN OLFACTOMETER *

Material	Number of Moths Caught	Average Catch per Replication	Percentage Ratio to Check
Methyl acetate...........	4	4	106.7
Methyl propionate........	11	11	293.3
Methyl n-butyrate........	14	14	373.3
Ethyl acetate............	2	2	53.3
Ethyl oxyhydrate.........	0	..	..
Ethyl propionate.........	6	6	160.0
Ethyl butyrate...........	5	5	133.3
Ethyl valerate...........	1	1	26.7
n-Propyl acetate..........	0	..	..
Isopropyl acetate........	1	1	26.7
Propyl propionate........	3	3	80.0
Isopropyl propionate......	0	..	..
Propyl butyrate..........	0	..	..
n-Butyl acetate..........	1	1	26.7
Isobutyl acetate.........	2	2	53.3
n-Butyl propionate........	1	1	26.7
Isobutyl propionate.......	2	2	53.3
n-Butyl n-butyrate........	1	1	26.7
Isobutyl butyrate........	0	..	..
Butyl valerate...........	2	2	53.3
Isobutyl valerate........	2	2	53.3
Amyl formate............	0	..	..
Isoamyl acetate.........	0	..	..
Isoamyl propionate.......	1	1	26.7
Isoamyl butyrate........	0	..	..
Amyl valerate...........	0	..	..
Amyl caproate..........	1	1	26.7
Check.................	15	3.75	100.0

* After Eyer and Medler, 1940.

$<\alpha$-toluic (240). Only a few other esters showed any pronounced degree of attractiveness (Table 19). Of these the attractiveness of esters of malonic, malic, and valeric acids is worthy of note because of the occurrence of the acids and their salts in the preferred food-plants, apple and pear.

Table 19

RESULTS OF COMPARISONS OF ESTERS OF HYDROXY, DIABASIC, AND AROMATIC ACIDS IN AN OLFACTOMETER *

Material	Number Caught	Average Catch per Replication	Percentage Ratio to Check
Methyl oxalate............	1	1	26.7
Methyl lactate............	5	5	133.3
Methyl malonate..........	6	6	160.0
Methyl benzoate..........	6	6	160.0
Methyl α-toluate..........	9	9	240.0
Methyl cinnamate.........	2	2	53.3
Ethyl oxalate.............	6	6	160.0
Ethyl lactate.............	0	0	..
Ethyl malonate...........	7	7	86.7
Ethyl succinate...........	0	0	..
Ethyl tartrate............	2	2	53.3
Ethyl citrate.............	2	2	53.3
Ethyl benzoate...........	0	0	..
Ethyl α-toluate...........	4	4	106.7
Phenyl ethyl phenyl acetate..	1	1	26.7
Phenyl ethyl phenyl cinnamate	1	1	26.7
Ethyl cinnamate..........	0	0	..
n-Propyl oxalate..........	0	0	..
n-Propyl lactate..........	2	2	53.3
Isopropyl oxalate.........	0	1	26.7
Isopropyl lactate.........	1	1	26.7
Isopropyl benzoate........	1	1	26.7
n-Butyl oxalate...........	0	0	..
n-Butyl lactate...........	2	2	53.3
Isobutyl lactate..........	0	0	..
n-Butyl dl-malate.........	5	5	133.3
n-Butyl tartrate..........	0	0	..
n-Butyl citrate...........	0	0	..
n-Butyl benzoate..........	0	0	..
Isobutyl benzoate.........	0	0	..
n-Butyl α-toluate.........	3	3	80.0
Isobutyl α-toluate........	2	2	53.3
Isoamyl oxalate..........	0	0	..
Isoamyl tartrate..........	0	0	..
Isoamyl benzoate.........	0	0	..
Benzyl cinnamate.........	1	1	26.7
Check...................	15	3.75	100.0

* After Eyer and Medler, 1940.

Field tests substantiated olfactometer tests. Acids and esters are more attractive than alcohols, the most attractive being α-toluic > butyric > acetic (Table 20). The best ethyl ester is ethyl α-toluate. When some of the more attractive acids were added to standard cane sirup bait, only citric and benzoic proved superior (Table 21).

Table 20

FIELD COMPARISON OF EMULSIONS OF CHEMICALS AS ATTRACTANTS FOR CODLING MOTHS DURING FLIGHT PERIOD OF THE OVERWINTERING GENERATION, APRIL–MAY, 1937 *

Material	Average per Trap per Day	Per Cent Total	Per Cent Ratio to Sirup Check
ALCOHOLS			
Ethyl.....................	1	2.38	33.3
n-Propyl...................	1	2.38	33.3
n-Butyl....................	1	2.38	33.3
Capryl....................	2	4.75	66.7
ACIDS			
Formic....................	1	2.38	33.3
Acetic....................	2.6	6.18	86.7
Propionic.................	2	4.75	66.7
Butyric...................	4	9.50	133.3
Lactic....................	2	4.75	66.7
Malonic...................	2.5	5.94	83.3
Citric....................	2	4.75	66.7
Benzoic...................	..	4.75	66.7
α-Toluic..................	7	16.63	233.3
ESTERS			
Ethyl propionate...........	1	2.38	33.3
Ethyl butyrate.............	1	2.38	33.3
Ethyl lactate..............	1	2.38	33.3
Ethyl succinate............	1	2.38	33.3
Ethyl tartrate.............	1	2.38	33.3
Ethylphenyl acetate........	3	7.13	100.0
n-Propyl acetate...........	1	2.38	33.3
CHECK			
Cane sirup...............	3	7.13	100.0

* After Eyer and Medler, 1940.

In general the most attractive acids are six-carbon and aromatic while the least attractive are aliphatic and simple dibasic. Of alcohols, the carbinols, especially phenyl ethyl carbinol, are superior in attractive qualities (Table 22). The best aldehyde tested was propionaldehyde (Table 23).

The correlation between gas production in sirup baits and moth catch deserves more study. Table 24 suggests that hydrogen and carbon dioxide are more than neutral carriers of aromas, in fact are mild attractants.

Table 21

AVERAGE CATCH AND ELIMINATION LEVEL FOR FOUR REPLICATIONS OF CANE SIRUP AND ATTRACTIVE CHEMICALS*

Bait Standard Cane Sirup Plus	Catch and Standard Error	Difference	Standard Error of Difference	Difference Divided by Standard Error of Difference
Sodium benzoate...	33 ± 3.38	..	..	..
Citric acid........	30 ± 2.08	3	3.97	0.7557
Benzoic acid.......	25 ± 1.22	8	3.59	2.2284
Baker's yeast.......	17 ± 1.39	16	3.65	4.3836
Tartaric acid......	17 ± .86	16	3.49	4.5845
Sirup alone........	15 ± .65	18	3.44	5.2326
Malic acid........	14 ± .74	19	3.46	5.4913
Malonic acid......	10 ± .65	23	3.44	6.6860
Oxalic acid.......	9 ± .41	24	3.40	7.0588
Butyric acid.......	4 ± .00	29	3.38	8.5799

* After Eyer and Medler, 1940.

Table 22

COMPARISONS OF ACETYL CARBINOL WITH RELATED ALCOHOLS*

Material	Olfactometer		Field Cage Average Catch per Day		Field (in Sirup)	
	Average Catch per Replication	Per Cent Ratio to Check	Emulsion	Sirup	Average Catch per Trap per Day	Per Cent Ratio to Sirup Check
Acetyl methyl carbinol.........	1.8	180	2.4	0.6	0.9	128.6
Isobutyl methyl carbinol.......	2.4	240	2.2	1.5	0.9	128.6
Phenyl methyl carbinol........	2.6	260	2.5	0.4	1.2	171.4
Phenyl ethyl carbinol..........	2.2	220	2.9	1.8	2.0	285.7
Sirup check..................	1.0	100	0.0	0.4	0.7	100.0

* After Eyer and Medler, 1940.

The work of Eyer and his associates with codling moths has, in short, established the following very important facts: (1) Adult codling moths are attracted to fermenting baits (and natural fermenting fruits or plant juices) by mixtures of compounds produced by the metabolic processes of a complex microflora of which certain species of bacteria are most important. (2) Of the compounds which may be produced, a wide variety of alcohols, acids, esters, and

Table 23

COMPARISON OF ACETIC ACID AND SOME RELATED ALDEHYDES AND KETONES IN FIELD CAGE*

Material	Average Catch per Day	Per Cent of Total	Per Cent Ratio to Emulsion Check
Acetic acid................	2	7.4	200
Propionic acid............:	1	3.7	100
Acetaldehyde..............	2	7.4	200
Propionaldehyde...........	4	14.8	400
Valeraldehyde.............	2	7.4	200
Acetone...................	3	11.1	300
Methylethyl ketone.........	3	11.1	300
Diethyl ketone.............	2	7.4	200
Emulsion check............	1	3.7	100
Cane-sirup check...........	7	25.9	700

* After Eyer and Medler, 1940.

Table 24

FIELD CAGE COMPARISONS OF GASES*

Gas	Average per Trap	Per Cent Total	Per Cent Ratio to Water Check
Oxygen (from sodium peroxide)...........	0	0	0
Oxygen (from hydrogen peroxide)........	0.71	19.3	200.9
Hydrogen (from Zn + HCl)..............	1.07	28.8	305.7
Carbon dioxide (from $CaCO_3$ + HCl)......	1.57	42.3	448.6
Water.................................	0.35	9.8	100.0

* After Eyer and Medler, 1940.

gases are attractive. These conclusions agree very nicely with the fact that the codling moth is rather rigidly restricted in its oviposition habits and less so in its feeding habits. Its ovipositional-type attractants appear to be characteristic essential oils while its feeding-type attractants are of more universal occurrence, some of them being found in practically any fermenting fruit or plant sap.

Hundreds of chemicals chosen more or less at random have been tested, either in pure form or mixed with fermenting baits, as attractants for the codling moth. The list includes, in addition to compounds that might reasonably be expected to occur in fermenting plant stock, numerous compounds found only in certain essential oils. Of those tested the following evince some degree of attractiveness: isobutyl phenyl acetate, diphenyl oxide, bromostyrol, benzyl benzoate, safrol (Yetter, 1925), oil of cloves, citronella, and sassafras (Yothers, 1927), ethyl propionate (Steiner, 1929), oils of bergamot, mace, and fennel, cymene, valeric acid, pine-tar-oil, nicotine sulfate, limonene, citral, ethyl p-toluidine (Van Leeuwen, 1935), and anethole (Bobb, Woodside, and Jefferson, 1939). Other than serving to produce a superior bait this information tells us very little from an absolute point of view. When the test substances are floated in evaporating dishes on standard bait, there is a mixing of odors. Van Leeuwen (1943) found that safrol, n-butyl sulfide, and pine-tar-oil presented in this manner caused an increase in catch of 300 per cent. An increase of 100 per cent was produced by citral, n-butyl disulfide, β-methylnaphthalene, geranyl acetate, fennel oil, bergamot oil, phenyl ethyl acetate, sassafras oil, oxidized kerosene, mace oil, benzyl ether, crude rosin spirits, limonene, anise seed oil, and ethylidene aniline.

Where the test substances are added to stock baits, any number of reactions resulting in the formation of any number of unknown compounds of untested attractive powers may be formed. Changing the substrate alters the products formed by fermentation. The uses of arsenic compounds, sulfites, and benzoates to increase or prolong the attractiveness of baits are cases in point. As Frost (1928) demonstrated, such compounds are not in themselves attractive. In specific tests sodium arsenite, benzoate, oleate, chloride, and hypochlorite were not attractive to the oriental peach moth. The action of such salts may appreciably increase the attractiveness as follows: (1) A high rate of fermentation occurs in the presence of a high phosphorus concentration; however, phosphorus is depleted by conversion into sugar phosphoric ester. By accelerating the regeneration of phosphates from esters, arsenates assist in the maintenance of rapid and continued fermentation. (2) Sodium sulfite increases acetaldehyde and glycerol production while decreasing that of carbon dioxide and alcohol. (3) Alkalis, especially carbonates, cause the breakdown of acetaldehyde to acetic acid and alcohol.

There are indications that some of the aromatic compounds listed above may have been performing only as ovipositional-type attractants. Of the moths attracted to oil of cloves, citronella, and sassafras 50 to 60 per cent were females of which 95 per cent were gravid (Yothers, 1927). Worthley and Nicholas (1937) found that anethole in a standard bait caused considerable increase in the number of females caught.

From the economic point of view several very successful baits have been developed. One formula calls for molasses 100 cc., water 900 cc., nicotine sulfate 1 cc., and pine-tar-oil 1 oz. (Van Leeuwen, 1939). The addition of valeric acid is optional. Another is composed of 1 part stock sirup, 20 parts water, 1 cc. anethole per quart, and sodium benzoate during the summer.

Oriental Fruit Moth Attractants. Shortly after Peterson (1925) first demonstrated the importance of fermentation in baits and Frost (1928) firmly established the fact by showing that oriental fruit moths would not respond to dry or highly concentrated unfermented mixtures of such sugars as lactose, levulose, maltose, arabinose, or dextrose, many papers treating of the attractiveness of various aromatic compounds to codling and oriental fruit moths began to appear. As in most of the codling moth studies the search for a perfect bait was in full cry, but the results were again comparative and not uncommonly inconclusive because substances to be tested were frequently mixed with stock sirup baits.

The oriental fruit moth, *Grapholitha molesta* Busck, in its natural habitat is subject to the same types of attractants as is the codling moth. It is to be expected that ovipositional-type attractants occur in essential oils and feeding-type attractants in fermenting plant juices. Of the hundreds of substances investigated as possible attractants no discrimination was made between the two types. Any attempt to evaluate studies on the oriental fruit moth should take these facts into consideration.

The most extensive contributions have been those of Frost (1928, 1935, 1936, 1936a, and 1937a), Yetter and Steiner (1931), Steiner and Yetter (1933), Worthley and Nicholas (1937), and Bobb (1938a). The results of Frost's extensive field experiments repeated over several seasons are extremely interesting. The different compounds tested included ketones, aldehydes, phenols, aromatic terpenes, esters, organic acids, and essential oils. Table 25 lists the leading materials for each season.

Table 25

BEST ATTRACTANTS FOR THE ORIENTAL FRUIT MOTH*

1933	1934	1935	1936
Methyl allephenol....	Linalool	Linolic acid	Oleic acid U.S.P.
Amyl acetate........	Safrol	Acetic acid	Terpinyl acetate
Citral..............	Propyl acetate	Anisic aldehyde	Safrol
Methyl cinnamate....	Amyl acetate	Oleic acid U.S.P.	Eugenol
Succinic acid	Anethole	Eugenol	Oleic acid (linolic free)
Soap	Fennel seed oil	Anisic acid	Oleic acid (Commercial)
Oleic acid..........	Terpinyl acetate	Oleic acid (linolic free)	Tartaric acid
Salicylic acid.......	Furfural	Anethole	Linolic acid
Sodium oleate.......	"	Safrol	Sodium oleate
Formaldehyde.......	"	Linseed oil	Acetone
Tartaric acid.......	"	"	"

* Data from Frost, 1935, 1936a, 1937a.

There is every reason to believe that some of the above-mentioned compounds are ovipositional-type and some feeding-type attractants. In view of Eyer's work it is probable that the acetates, acetone, and acetic acid belong to the latter category. Linalool, safrol, furfural, anethole, eugenol, anisic acid, and anisic aldehyde are characteristic of essential oils and probably operate in the former capacity. Frost's observation that specific materials are attractive at specific periods of the season lends support to this belief. Terpinyl acetate is most effective during May and June. The acids, tartaric excepted, are best from August to September, toward the end of the season when oviposition becomes the prime activity.

The great similarity in habits of the codling moth and oriental fruit moth would lead one to expect that the same substances would be attractive to both. In many respects this is so, the differences between the two being little greater than the seasonal differences for either one alone. Some of the substances most attractive to codling moths have not been tested with the oriental fruit moth and vice versa so that strict comparisons are not possible. Isophenylbutyl acetate and ethyl oxyhydrate, for example, tested by Eyer, *et al.*, should be tried with oriental fruit moths. Generally speaking, however, acetates rank high with both. Methyl cinnamate likewise is a favorite. It is curious that safrol, fennel oil, and anise compounds (i.e., anisic aldehyde, anisic acid, anethole) should be so attractive to both species. As is known, these compounds are characteristic of the essential oils of Umbelliferae and are either absent or present in minute amounts only in fruit trees.

Attractants of Cabbage Looper Moths. Cabbage loopers are the larvae of moths of the genus *Autographa*, some of the most frequent visitors to moth baits, tree wounds exuding sap, and heavily scented flowers. Smith, Allen and Nelson (1943) have recently investigated the possibility of baits for use against these pests. Their tests with *A. brassicae*, *A. oo* (Cram.), *A. oxygramma* (Geyer), *A. ou* (Guen.), *A. verruca* (F.), *A. basigera* (Walk.), and *A. biloba* (Steph.) indicated that diphenyl ether, benzyl valerate, benzyl ether, phenylacetaldehyde, and palmitaldehyde are effective attractants. When very pure samples were fractionated, none of the fractions was attractive. In the benzyl acetate series the attractive principle is an unknown foreign impurity; in the phenylacetaldehyde series, either an impurity or the highest fraction.

It is obvious that these are feeding-type attractants and normal constituents of fermenting carbohydrates in nature. Ovipositional-type attractants would be mustard oils. It would be of interest to determine whether or not gravid females could be caught in traps baited with mustard oils.

Fruit Fly Attractants. Of the multitudes of insects which attack green, ripe, or fermenting fruits the economically important fruit flies, especially those comprising the genera *Drosophila*, *Pterandrus* (*Ceratitis*), *Rhagoletis*, *Anastrepha*, *Dacus*, and *Chaetodacus* have been the subject of considerable investigation. These belong for the most part to the Drosophilidae (pomace flies) and the Trypetidae (fruit flies). The former, as adults, feed chiefly on fermenting fruits. As larvae they breed in decaying fruit, fungi, or even beneath the epidermis of leaves as miners. Adults of the latter family feed more commonly on dew and exudations of plant sap although they too are to some extent attracted to fermenting carbohydrates. In the larval stage they bore into living fruits and stems.

It has long been common knowledge that *Drosophila* is strongly positively chemotactic to fermenting banana. Banana contains 16 per cent fermentable sugars from which is derived ethyl alcohol to the extent of 6.55 to 10.12 per cent. Continued fermentation produces 5.72 per cent acetic acid (Von Loesecke, 1929). In addition to these two compounds fermenting fruit also contains amyl alcohol, butyric acid, lactic acid, and acetic ether. In his early work Barrows determined that the most attractive concentrations of ethyl alcohol and acetic acid were 20 per cent and 5 per cent respectively while a mixture containing 2.5 per cent alcohol and 0.625 per cent acetic acid together proved even more attractive. On the basis of these data and those of Von Loesecke, Reed (1938) studied extensively the reactions of *Drosophila melanogaster* to solutions of ethyl alcohol ranging from 5 to 20 per cent and of acetic acid ranging from 0.1 to 5 per cent. These experiments revealed several facts regarding the profound effects of concentration (cf. Chap. 6) on the responses of insects to chemicals. In brief, females of this species respond to solutions of ethyl alcohol up to 15 per cent, with a maximum at or below 5 per cent. Acetic acid is attractive to both sexes in concentrations not exceeding 1 per cent. The maximum response for females is at 0.4 per cent; for males, 0.2 per cent. Above 25 and 5 per cent respectively ethyl alcohol and acetic acid behave as repellents. Hutner, Kaplan and Enzmann (1937) found that diacetyl, acetyl methyl carbinol, dioxane, acetyl cyclohexane, diphenyl methane, and β-bromethyl acetate also are powerful attractants. Reactions to the first two chemicals are clearly understandable; diacetyl (CH_3—CO—CO—CH_3) is partially responsible for the odor of yeast, and acetyl methyl carbinol rapidly oxidizes to diacetyl. The remaining compounds exemplify that category of substances attractive to insects but not known to occur in nature. A less decisive response follows stimulation by lower aliphatic amines, acetaldehyde, and indole.

Fermentation products are also attractive to certain Trypetidae (cf. p. 119). Ripley and Hepburn (1929, 1929a, 1931) conducted an exhaustive series of experiments in which they tested the responses of the Natal fruit fly, *Pterandrus rosa* (Ksh.), both in the field and laboratory to several hundred substances. In order of decreasing effectiveness the best attractants are fermenting pollard bait, oil of cloves, and linalyl acetate (a fruit ester from peaches, the optimum concentration of which is 0.1 to 0.15 per cent) (cf. p. 24). In general, the most powerful attractants are aliphatic alcohols, aldehydes, esters, and acids with few carbon atoms. This is quite in contrast to the codling moth for which aliphatic compounds rank low as attractants while acids and esters, especially those that are aromatic and contain six carbon atoms, rank high. In view of this information and the fact that there is a cyclic production of compounds in fermenting baits, it is interesting to observe that there is a correlation between the age of a given bait and the species of insects attracted.

In a study of Coleoptera taken from bait traps Frost and Dietrich (1929) noted three periods in molasses bait fermentation: the first, lasting from one to two days, is characterized by alcoholic fermentation; the second, of three to four days duration, is a period of acetic fermentation; the third is a period of putrefaction. During period one and the early part of period two Nitidulidae and Scarabaeidae, which normally feed on flowers, arrive in numbers. Trypetidae also arrive during period one; Drosophilidae, during period two. During period three Silphidae and a few Scarabaeidae predominate. To putrid baits also came a few Histeridae and occasional Helodidae, Dermestidae, and Byrrhidae.

Recently Starr and Shaw (1944) have discovered that pyridine in alcoholic solution increases the attractiveness of fermenting baits for *Anastrepha ludens*. Experiments in which alcohol was omitted indicated that pyridine was the active principle. The action of this compound is not clear.

Fermentation Products Attractive to Xylophagous Insects. Probably because of the clandestine and unobtrusive activities of borers the feeding habits of these insects are but little understood. Different species bore into living trees, dying trees, dead trees, and rotten wood; they feed on the sap, the wood, or the associated fungi. The confined and sheltered mode of existence of some, moreover, obviates the necessity of chemical feeding-type attractants. Ovipositional-type attractants seem to be more important. Whereas some species, e.g., Cerambycidae and Buprestidae, oviposit on the outer bark, others, notably the Scolytidae, oviposit at the end of borings. When attractants do perchance direct feeding, they are acting in the dual rôle of feeding-type and ovipositional-type attractants.

In the absence of direct experimental work the hypothesis of Person (1931) regarding feeding-type attractants of the bark beetles *Ips pini* and *I. grandicollis* Eich. deserves mention, especially since it suggests also how some strictly mycetophagous insects may be attracted to the products of microfloral action on a substratum. In this case it is believed that yeasts associated with the bark beetles initiate fermentation in the sap of infested trees with the resultant formation of odors attractive to other beetles. Thus infestation by large numbers of beetles is brought about. In view of the token nature of odors it is worthy of emphasis that the fungus instrumental in the production of the odors is distinct from that upon which the

FIG. 18. Collar method of injecting chemicals into living trees. (Courtesy, U.S. Department of Agriculture, Bureau of Entomology and Plant Quarantine.)

beetles feed. The observation of Buchanan (1941) that the ambrosia beetle, *Xylosandrus germanus* (Blfd.) is especially attracted to living elm trees into which ethyl alcohol has been injected now takes on new meaning and lends weight to Person's hypothesis. Further corroboration is to be found in the knowledge that species of Nitidulidae, especially *Ips fasciatus*, are very strongly attracted by fermenting molasses.

Extensive use of the injection of chemicals into living trees as a means of preserving wood is based on an idea many centuries old. As a procedure in the study of attractants it is a relatively new technique that should yield additional valuable information. The most practical method of introducing liquid chemicals into living trees consists in girdling the xylem and tacking a collar of rubberized fabric around and joined to the tree just below the girdle (Fig. 18). After the base of the collar has been sealed with an asphaltum compound, the chemical solution is poured in (Craighead and St. George, 1938; Lantz, 1938; Whitten, 1941; Wilford, 1944). Elms thus treated and killed by sodium chlorate prove to be exceptionally attractive to the bark beetles *Scolytus multistriatus* Marsham and *Hylurgopinus rufipes* (Eich.) (Whitten and Baker, 1939). Assuredly this is a stimulating problem meriting further attention.

Discussion. Generally speaking, insects attracted by the products of fermentation are not rigidly specific in their feeding habits. As with other substances, the odor of fermenting substrates results from the blending of many compounds. The available compounds, however, are limited in number as compared with those in essential oils. Thus there is a sameness about the odors of fermenting plant products as contrasted with the numerous characteristic odors of living plants that does not make for specificity among the insects feeding on such substances. This is one of the reasons that multitudes of wood-boring, sap-feeding, honey-sucking, and fruit-feeding insects are attracted to the same baits. That there is a degree of specificity, however, is attested by the fact that some species are attracted in greater numbers than others. Also, certain insects prefer baits of a given composition, this preference undoubtedly being governed by the characteristic fermentation products resulting from action on different kinds of substrates. Thus, honey bees prefer water and sugar solutions; Syrphidae, low grade molasses. Furthermore, among the limited compounds available in a fermenting mixture certain ones in pure form are more attractive to some insects than to others. This might be explained by assuming that under natural conditions all of the insects attracted to the same bait may not necessarily be responding to the same component of that bait, that insects are able to analyze the mixtures and detect one of the odors, and that by adding chemicals to stock baits we simply increase the concentration of a preferred odor. Among preferences noted is that of Noctuidae for baits to which have been added citric, malic, and tartaric acids; Ortalidae for baits containing amyl acetate, acetic and malic acids; Tabanidae for soap, sodium oleate, oleic acid, camphoric acid; Chrysopidae for amyl acetate, pinene, citrene.

The annual alternation in popularity of one compound over

another, most strikingly observed in studies of oriental fruit moths and cabbage loopers, is another example, at the moment inexplicable, of specificity.

Probably here more than elsewhere factors other than chemical greatly influence the feeding habits of an insect so that the relative abundance of species at baits does not represent true ratios. As Frost noticed, baits in wooded areas attracted more Cerambycidae than identical ones placed elsewhere. Also, the amount of sun and humidity is of prime importance. Middleton (1929) has pointed out other less obvious factors of ecological significance that are pertinent here. He noted that infestation by shade tree insects is a function of rainfall, physiological condition of the tree, and environmental position in regard to shade, density of stand, etc.

A perusal of lists of insects attracted to fermenting baits or the products thereof also impresses one with the fact that many insects are lured that do not normally, according to the best of our knowledge, have any relation to natural fermenting materials. Thus, relatively large numbers of female tabanids, which are normally blood-feeders, have been taken. Predatory ambush bugs (Phymatidae) occur in large numbers. There is always a remote possibility that such insects may be attracted by odors that attract their prey. Such could also be true of parasitic species. The behavior of *Pimpla ruficollis*, a parasite of the pine shoot moth, *Rhyacionia buoliana* Schiff., toward the oil of *Pinus sylvestris* has been shown by Thorpe and Caudle (1938) to be an association of this sort. As a rule, though, very few parasites are taken at baits. *Oncodes incultus* (O.S.), a parasite of spiders, is an exception. Some products of fermentation apparently are generally attractive to many insects that accept them in passing, so to speak. This is probably true of tree hoppers (Membracidae) and leaf hoppers (Jassidae) which are very constant visitors.

Of those insects that might reasonably be expected to come to fermenting baits because of the nature of their food, carpophagous species are represented by the oriental fruit moth, codling moth, bumble flower beetle (*Euphoria inda:* Scarabaeidae), some Trypetidae, and many Drosophilidae. For some reason the plum curculio is not attracted, probably because it attacks green and ripe and not fermenting fruit. Of the Coleoptera, which have been extensively studied, at least 40 families are found to be frequent visitors to fermenting baits. Most conspicuous in number of genera and species are Carabidae, Staphylinidae, Elateridae, Cerambycidae, Chrysomelidae (accidental catches), Curculionidae, and Scolytidae. Most numerous in numbers of individuals are Staphylinidae, Nitidulidae, and Cerambycidae. Many of the Staphylinidae taken normally feed on fermenting sap; the majority of Nitidulidae, on fermenting plant juices and sap. Of xylophagous insects Cerambycidae assemble in remarkable numbers and Scolytidae in relatively large numbers. Buprestidae are strikingly uncommon (see p. 33). Peach borers are attracted to essential oils only. Curiously enough, honey-feeders are among the missing. Stranger still, aromatic odors likewise fail to attract this group, but this failure might be explained by the choice of substances tried. Honey bees and Syrphidae are infrequent visitors to sweet and aromatic baits, those with citral or anethole excepted.

A sidelight of these observations is the appearance of various insects whose feeding habits are little understood. Among those commonly taken are Hemerobiidae and Chrysopidae. The latter arrive in enormous numbers.

Finally, it may be said that the products of fermentation function as feeding-type attractants for innumerable insects and ovipositional-type attractants for a few like the Drosophilidae and Scolytidae. Of the multitudes that are attracted for food most resort to the components of essential oils as guides when seeking ovipositional sites.

REFERENCES

Bobb, M. L.: The use of bait traps in codling moth control, *Virginia Fruit*, **26**(4), 20–24 (1938).

——: Bait traps for the control of the oriental peach moth, *Virginia Agr. Exp. Sta. Bull*. 314, 1–14 (1938a).

——, A. M. Woodside, and R. N. Jefferson: Bait and bait traps in codling moth control, *Ibid.*, 320, 1–19 (1939).

Buchanan, R. E., and E. I. Fulmer: "Physiology and Biochemistry of Bacteria," Baltimore, The Williams and Wilkins Co., 1930, v. 3.

Buchanan, W. D.: Experiments with an Ambrosia Beetle *Xylosandrus germanus* (Blfd.), *J. Econ. Entomol.*, **34**(3), 367–369 (1941).

Champlain, A. B., and H. B. Kirk: Bait pan insects, *Entomol. News*, **37**(9), 288–291 (1926).

Craighead, F. C., and R. A. St. George: Experimental work with the introduction of chemicals into the sap stream of trees for the control of insects, *J. Forestry*, **36**(1), 26–34 (1938).

Cuscianna, N.: Osservazioni sull'attrazione esercitata degli odori sugli insetti, *Bol. lab. zool. gen. e agrar.*, *R. scuola sup. d'agric. Portici*, **15**, 226–253 (1922).

Eyer, J. R.: A four-year study of codling moth baits in New Mexico, *J. Econ. Entomol.*, **24**(5), 998–1001 (1931).

——: Further observations on limiting factors in codling moth bait and light trap attrahency, *Ibid.*, **27**(3), 722–723 (1934).

——: Ten years' experiments with codling moth bait traps, light traps and trap bands, *New Mexico Bull.* 253 (1937).

——, and J. T. Medler. Attractiveness to codling moth of substances related to those elaborated by heterofermentative bacteria in baits, *J. Econ. Entomol.*, **33**(6), 933–940 (1940).

——, ——, and H. P. Linton: Analysis of attrahent factors in fermenting baits used for codling moth, *Ibid.*, **30**(5), 750–756 (1937).

——, and H. Rhodes: Preliminary notes on the chemistry of codling moth baits, *Ibid.*, **24**(3), 702–711 (1931).

Frost, S. W.: Beneficial insects trapped in bait-pails, *Entomol. News*, **38**, 153–156 (1927).

——: Continued studies of baits for oriental fruit moth, *J. Econ. Entomol.*, **21** (2), 339–348 (1928).

——: Notes on Pennsylvania Ortalidae (Dipt.), *Entomol. News*, **40**(3), 84–87 (1929).

——: 1934 notes on baits for oriental fruit moth, *J. Econ. Entomol.*, **28**(2), 366–369 (1935).

——: A summary of insects attracted to liquid baits, *Entomol. News*, **47**(3), 64–68, 89–92 (1936).

——: Tests on baits for oriental fruit moth, 1935. *J. Econ. Entomol*, **29**(4), 757–760 (1936a).

——: New records from bait traps (Dipt., Coleop., Corrodentia), *Entomol. News*, **48**(7), 201–202 (1937).

——: Tests on baits for oriental fruit moths, 1936. *J. Econ. Entomol.*, **30**(5), 693–695 (1937a).

——, and H. Dietrich: Coleoptera taken from bait-traps, *Ann. Entomol. Soc. Am.* **22**(3), 427–437 (1929)

Garcia, F.: Fifty-second Annual Report, New Mexico Agr. Exp. Sta., **1941**, 55–58.

Harrow, B., and C. P. Sherwin: "A Textbook of Biochemistry," Philadelphia, W. B. Saunders Co., 1935.

Hettche, H. O., and B. Weber: Die Ursache der bakteriziden Wirkung von Mesentericusfiltraten, *Arch. Hyg.*, **123**, 69–80 (1939).

Hutner, S. H., H. M. Kaplan, and E. V. Enzmann: Chemicals attracting *Drosophila*, *Amer. Naturalist*, **71**, 575–581 (1937).

Lantz, A. E.: An efficient method for introducing liquid chemicals into living trees, *U.S. Dep. Agr., Bur. Entomol. Plant Quarantine*, E-434 (1938).

Luftensteiner, H.: Aetherische Oele und Geruchträger, *Pharm. Post*, **44**(70–72, 74–78), 711–714, 719–721, 727–730, 751–753, 767–768, 779–782, 791–794, 799–802 (1911).

Middleton, W.: Factors influencing the activity of shade-tree insects and the utilization of these in control work. *Trans. 4th Internat. Cong. Entomologists, Ithaca*, 1928, **2**, 374–381 (1929).

Person, H. L.: Theory in explanation of the selection of certain trees by the western pine beetle, *J. Forestry*, **29**, 696–699 (1931).

Person, A.: A bait which attracts the oriental peach moth (*Laspeyresia molesta* Busck), *J. Econ. Entomol.*, **18**(1), 181–190 (1925).

Reed, M. R.: The olfactory reactions of *Drosophila melanogaster* Meigen to the products of fermenting banana, *Physiol. Zoöl.*, **11**(3), 317–325 (1938).

Ripley, L. B., and G. A. Hepburn: Olfactory and visual reactions of the Natal fruit-fly, *Pterandrus rosa* (Ksh.), as applied to control, *S. African J. Sci.*, **26**, 449–458 (1929).

——, and ——: Studies on reactions of the Natal fruit-fly to fermenting baits, *Union S. Africa Dep. Agr., Entomol. Mem.*, **6**, 19–53 (1929a).

——, and ——: A new olfactometer successfully used with fruit-flies, *Ibid.*, **6**, 55–74 (1929b).

——, and ——: Further studies on the olfactory reactions of the Natal fruit-fly, *Pterandrus rosa* Ksh., *Ibid.*, **7**, 24–81 (1931).

Smith, C. E., N. Allen and O. A. Nelson: Some chemotropic studies with *Autographa* spp., *J. Econ. Entomol.*, **36**(4), 619–621 (1943).

Starr, D. F.: A mixture of pyridine and iso-amyl alcohol attractive to some species of *Blepharoneura*, *Ibid.*, **37**(4), 547 (1944).

——: and J. G. Shaw: Pyridine as an attractant for the Mexican fruit-fly, *Ibid.*, **37**(6), 760–763 (1944).

Steiner, L. F.: Codling moth bait trap studies, *Ibid.*, **22**(4), 636–648 (1929).

——, and W. P. Yetter: Second report on the efficiency of bait traps for the oriental fruit moth as indicated by the release and capture of marked adults, *Ibid.*, **26**(4), 774–788 (1933).

Thorpe, W. H., and H. B. Caudle: A study of the olfactory responses of insect parasites to the food plant of their host, *Parasitology*, **30**(4), 523–528 (1938).

Van Leeuwen, E. R.: Investigations of baits attractive to the codling moth, *Proc. 31st Ann. Meeting Wash. State Hort. Assoc.*, **1935**, 136–139.

——: Further contributions to a study of baits for the codling moth, *Proc. 35th. Ann. Meeting Wash. State Hort. Assoc.*, **1939**, 21–29.

——: Chemotropic tests of materials added to standard codling moth bait, *J. Econ. Entomol.*, **36**(3), 430–434 (1943).

Von Loesecke, H.: Preparation of banana vinegar, *Ind. Eng. Chem.*, **21**(2), 175–176 (1929).

Whitten, R. R.: The internal application of chemicals to kill elm trees and prevent bark-beetle attack, *U.S. Dep. Agr., Circ.* 605, 1–12 (1941).

——, and W. C. Baker: Tests with various elm-wood traps for bark beetles, *J. Econ. Entomol.*, **32**(5), 630–634 (1939).

Wilford, B. H.: Chemical impregnation of trees and poles for wood preservation, *U.S. Dep. Agr., Circ.* 717, 1–30 (1944).

Worthley, H. N., and J. E. Nicholas: Tests with bait and light to trap codling moth, *J. Econ. Entomol.*, **30**(3), 417–422 (1937).

Yetter, W. P.: Codling moth work in Mesa County, *16th Ann. Rep. State Entomol., Colorado, 1924, Circ.* 47, 32–40 (1925).

——, and L. F. Steiner: A preliminary report on large-scale bait trapping for the oriental fruit moth in Indiana and Georgia, *J. Econ. Entomol.*, **24**(6), 1181–1197 (1931).

Yothers, M. A.: Summary of three years' tests of bait traps for capturing the codling moth, *Ibid.*, **20**(4), 567–575 (1927).

——: Summary of results obtained with trap baits in capturing the codling moth in 1927, *Ibid.*, **23**(3), 576–587 (1930).

Chapter 5

Protein and Fat Decomposition Products (Fatty Acids, Amines, Ammonia, Carbon Dioxide)

Source of Attractants. Proteins, fats, and oils are the great reserve materials of life. Together they primarily are responsible for mass in living matter. In such a capacity they are omnipresent wherefore they are to be reckoned with as the mainstay of all food. There is a bewildering array of these materials and numbered among them are the most complex of compounds.

Proteins and fats as such are usually odorless as a result of which they are incapable of acting as guiding attractants or repellents. The products of their partial or complete decomposition are, however, notably odorous, a character which bestows upon them the properties of attractiveness or repellence. In order to understand and examine the effects of these on insects it is expedient to enumerate the parent compounds. Those which concern us may perhaps be most conveniently grouped as follows:

Proteins
Simple proteins (composed entirely of amino acids)
Albumins ⎫
Globulins ⎭ (soluble)
Scleroproteins (insoluble)
Conjugated proteins (simple proteins plus a nonprotein group)
Phosphoproteins (contain phosphorus)
Glycoproteins (contain the carbohydrate radical)
Hemoglobins (contain hematin)
Nucleoproteins (contain nucleic acid)
Lipoproteins (contain fatty acids, phospholipins, etc.)
Lipins (also termed fats, lipids, lipoids)
True fats (esters of fatty acids and glycerol—no odor or color)

Phosphatides (phospholipins) (contain glycerol, fatty acids, phosphoric acid, and organic bases)

Neutral fats (mixed glycerides)

Fatty acids (usually volatile saturated monobasic aliphatic acids of a homologous series $(C_nH_{2n}O_2)$; may also embrace unsaturated acids)

Cerebrosides (compounds of fatty acids containing neither glycerol nor phosphoric acid)

Sterols (complex waxlike substances, monohydroxy alcohols)

Soaps (salts of higher fatty acids)

Common to all proteins is the ability to form alpha amino acids by hydrolysis, this reaction being initiated in nature by an enzyme or in the laboratory by boiling with acids or alkalis. In the former class of reactions the most important agents of decomposition are

Table 26

REACTIONS TAKING PLACE DURING THE DECOMPOSITION OF PROTEINS AND THE TYPES OF PRODUCTS FORMED *

Reaction	Products
1. Decarboxylation.	Amines
2. Hydrolytic deamination.	Hydroxy acids
3. Hydrolytic deamination and decarboxylation.	Alcohols
4. Reductive deamination.	Saturated acids
5. Reductive deamination and decarboxylation.	Hydrocarbons
6. Deamination.	Unsaturated acids
7. Oxidative breakdown	Compounds with few C atoms

* Data from Harrow and Sherwin: "A Textbook of Biochemistry," Philadelphia, W. B. Saunders Co., 1935.

Table 27

SPECIFIC EXAMPLES OF THE REACTIONS LISTED IN TABLE 26 *

	Type of Reaction	Organism	Action
Glycine.	1	Mixed	Putrefaction of fish
Leucine.	1	Mixed	Putrefaction of meat
Phenylalanine.	1	Mixed	Putrefaction of gelatin and meat
Phenylalanine.	4 and 7	Mixed	Putrefaction of meat and keratin
Tryptophane.	7	Mixed	Putrefaction of proteins and ox brain
Tyrosine.	1	Mixed	Putrefaction of proteins and meat

* Data from Harrow and Sherwin. "A Textbook of Biochemistry," Philadelphia, W. B. Saunders Co., 1935.

bacteria. Degradation by this microflora involves, in addition to cleavage to amino acids, reduction, oxidation, decarboxylation, deamination, etc. A list of the types of products to be expected from such reactions is given in Table 26.

Table 28

VARIOUS CHANGES WHICH FATTY SUBSTANCES MAY UNDERGO IN NATURE*

Substrate	Action	Products
Unsaturated fatty acids	Reduction	Corresponding saturated acids
Unsaturated fatty acids	Mild oxidation	Hydroxy acids (e.g., lactic, β-hydroxybutyric)
Unsaturated fatty acids	Vigorous oxidation	Mono- and dibasic-acids
Neutral fats	Hydrolysis (saponification) by heating, enzymes, or alkalis	Fatty acids and glycerol
Glycerides	Partial spontaneous hydrolysis (rancidity)	Free volatile odoriferous fatty acids
Fats and oils containing unsaturated acids	Oxidation (rancidity)	Aldehydes, ketones, and lower acids
Lecithins	Oxidation in air	Fatty acids with disagreeable odors
Phospholipins	Various	Glycerol, fatty acids, phosphoric acid, and organic bases

* Data from Harrow and Sherwin: "A Textbook of Biochemistry," Philadelphia, W. B. Saunders Co., 1935.

Fatty substances also undergo various changes in nature, some of which are listed in Table 28. By far the most important products are fatty acids. They may be classified as follows:

Saturated

1. Straight chain acids $(C_nH_{2n}O_2)$

acetic	$C_2H_4O_2$	in vinegar
butyric	$C_4H_8O_2$	in milk fat
caproic	$C_6H_{12}O_2$	
caprylic	$C_8H_{16}O_2$	in palm oils, etc.
capric	$C_{10}H_{20}O_2$	
lauric	$C_{12}H_{24}O_2$	in laurel
myristic	$C_{14}H_{28}O_2$	
palmitic	$C_{16}H_{32}O_2$	in animal and vegetable oils
stearic	$C_{18}H_{36}O_2$	

2. Hydroxy acids (e.g. lactic, $C_3H_6O_3$)

Unsaturated

1. Series with one double bond, $C_nH_{2n-2}O_2$ (e.g., oleic or acrylic—$C_{18}H_{34}O_2$—found mostly in fish oils, etc.)
2. Linolic series, $C_nH_{2n-4}O_2$ (e.g., all C_{18} series, found mostly in vegetable oils)
3. Linolenic series, $C_nH_{2n-6}O_2$

Occurrence of These Attractants in Nature. Although protein and fat decomposition products are widespread in nature, there are certain substances in which they exist in more or less characteristic aggregates where they behave as especially powerful attractants. Outstanding among them are carrion, feces, urine, animal secretions such as sweat, decomposing plant material, fungi, and a few algae. Coming to feed or oviposit on these are hosts of insects of more or less nonspecific feeding habits. Saprophagous species (scavengers) account for a large proportion of the group which includes such dung and carrion feeders as the Silphidae, Histeridae, Staphylinidae, Scarabaeidae, Muscidae, Sarcophagidae, and Calliphoridae; such aquatic scavengers as Dytiscidae and Gyrinidae; such geophagous members as crane-fly larvae; detrivorous families as Dermestidae and Tineidae; species that attack decaying fruit and other plant tissue; such entomophagous scavengers as ants. The remainder of the assemblage is made up of mycetophagous, phycophagous, entomophagous (parasitic), and harpactophagous (predatory) species.

Coprophagous and necrophagous species are attracted by the odorous constituents of decomposing flesh, feces, urine, and vomitus to the vicinity of which they first arrive by means of klino-kinetic movements. As the concentration gradient becomes steeper, they advance directly toward the source of odor. *Scarabaeus*, for example, travels straight to horse dung from a distance of 10 m.; *Geotrupes*, from a distance of 50 cm. A glance at Tables 29 and 30 reveals the chemicals in these substances which are available as attractants. The more active of these from a chemotactic point of view are indole, skatole, paracresol, the fatty acids, the mercaptans, the amines, ammonia, hydrogen sulfide, and carbon dioxide.

Entomophagous insects, ectoparasites such as lice, bedbugs, fleas, and ticks, and blood-sucking Diptera locate their hosts by a variety of signs. In the case of wingless forms the host must first come to the

Table 29

AVERAGE COMPOSITION OF HUMAN URINE AS COMPARED WITH THAT OF BLOOD PLASMA *

Compound	Urine	Blood Plasma
Urea	2%	0.033%
Other N compounds (uric acid, hippuric acid, ammonia, ammonium hydroxide, ammonium carbonate)	0.3	0.01
Sulfate	0.18	0.003
Phosphate	0.18	0.009
Sodium chloride	1.0	0.65
Salts	0.3	0.14
Glucose	trace	0.1
Amino acids	trace	0.05
Total crystalloids	4	1
Colloids (proteins and lipoids)	0	7
Water	96	92

* Data reprinted from "Outlines of Modern Biology," by C. R. Plunkett. Copyright, 1929, by Henry Holt and Company, Inc.

Table 30

COMPOSITION OF FECES WITH SPECIAL REFERENCE TO
ODOROUS CONSTITUENTS

Undigested material
Remains of dead intestinal cells
Remains of digestive secretions
Waste material secreted in bile
Dead bacteria (as much as 50% of total)
Bile pigments (giving the color)
Indole
Skatole
Paracresol
Phenol
Para-oxyphenylpropionic acid
Para-oxyphenylacetic acid
Volatile fatty acids
Benzoic acid
Indol-acetic acid
Methyl mercaptan
Some unaltered amino acids
Primary, secondary, and tertiary amines
Phenylethylamine
Indol-ethylamine
Diamines
    Putrescine (tetramethylene diamine)
    Cadaverine (pentamethylene diamine)
Hydrogen sulfide
Methane
Carbon dioxide, ammonia, proteoses, peptoses, peptides

Table 31

COMPOSITION OF ECCRINE SWEAT

Compounds	Concentration mg./100 cc.
Lactic acid....	82.05–160.4
Uric acid......	.0.017
Urea.........	0.16–0.28
Ammonia.....	5–35
Amino acids...	1.57–4.76
Sugar........	5.6–40
Fats (many)...	..

Table 32

COMPOSITION OF HUMAN SKIN

Keratin (composed of the three amino acids: glycine, glutamic acid, and cystine)

Collagen (amino acids)

Elastin

Neutral fat

Cholesterol esters

Cholesterol

Lipoids

Sugar

Table 33

COMPOSITION OF NORMAL BLOOD *

Substance	grams/100 cc.

PLASMA COMPOSITION

Water..	90.0
Proteins { Serum albumin........................	4.4
Serum globulin.......................	2.3
Fibrinogen...........................	0.3
Nonprotein nitrogen { Urea / Uric acid / Creatinine / Amino acids / Ammonia }	0.036
Glucose (apparent)...........................	0.100
Chlorine.....................................	0.36
HCO₃..	0.18
Phosphorus (inorganic).......................	0.003
Phosphorus (other compounds).................	0.020
Sodium......................................	0.335
Calcium.....................................	0.01
Potassium...................................	0.02
Magnesium..................................	0.003

RED CELL COMPOSITION

Water.......................................	60.0
Hemoglobin.................................	32.0
Chlorine....................................	0.18
HCO₃.......................................	0.09
Potassium...................................	0.42

Also gases, lactates, phenols, sterols, fats, soaps, phosphatides, and lactic acid

* After Bard: "MacLeod's Physiology in Modern Medicine," St. Louis, C. V. Mosby Co., 1938.

locale of the waiting parasite. This having been accomplished, the insect or tick may be guided by two types of stimuli, thermal and chemical. With winged blood-sucking forms vision may be of extreme importance in guiding them to their hosts. This is especially true of tsetse flies. Host odor, however, is still a factor. Screens scented with alcohol and ether extracts from the body surface of living bush-pigs, oxen, or sheep are significantly more attractive than unscented screens (Vanderplank, 1944).

Host odors are due principally to fresh and decaying body secretions, clotting and dried blood, and putrefying flesh in the vicinity of wounds. Some animals possess special glands producing strong characteristic odorous secretions. Of most common occurrence as sources of vertebrate host attractants are blood (Table 33), sweat (Table 31), and dead tissue. In addition to proteins, blood and tissues contain a large amount of fermentable sugar. In man, 12 hours after eating, there may be 80 mg. of sugar per 100 cc. of blood. Sugar concentration is higher in birds, lower in ruminants, and lowest in cold-blooded vertebrates. Sweat is a superb source of host odor. Its composition and odor varies as the location of the glands on the body. Sweat from the brow contains a high concentration of uric acid; that of the hand, chlorides; that of the thigh, lactate (Mickelson and Keys, 1943). Generally speaking, sweat is an aqueous secretion containing sodium chloride, urea, fats, and fatty acids. In man there are two kinds of sweat glands and two kinds of sweat, those connected with hair follicles and secreting oil, and those connected with pores and secreting perspiration. Eccrine sweat, secreted throughout the skin, is acid; apocrine, secreted in limited areas, e.g., groin, axilla, etc., is alkaline. The normal composition of eccrine sweat is given in Table 31. Here the chief odorous compounds are ammonia and fatty acids.

Some Diptera, as the stable fly *Stomoxys calcitrans*, are attracted by perspiration and other host odors. According to Marchand (1918), however, species of *Culex* ignore perspiration and fresh blood but will attempt to pierce an artificial object heated to 98.6° F. Actually, the voluminous work directed toward solving the problem of host selection by this group of insects has produced meager results. For example, no complete explanation of the factors guiding mosquitoes to blood meals has been presented. Some of the materials tested and found unattractive are: urine, urea, glycocoll, cystine, phenylalanine, aspartic acid, asparagine, glutaminic acid, alanine, trypsin, tyrosin, gliodin, various concentrations of ammonium hydroxide, ammonium carbonate, acetic acid, sweat, human sebaceous secretions, cholesterol, oleic acid, benzoic acid, a mixture of benzoic and oleic acids, a mixture of oleic acid and cholesterol, stearic acid, blood of various animals, pepsin, fermented dextrose, peptone, sugar solutions, valeric acid, lactic acid, indole, skatole, etc. (Rudolfs, 1922). According to this worker, substances which exhibit any appreciable powers of attraction are combinations of carbon dioxide, ammonia, humidity, and temperature such as exist in human breath. Davidson, *et al.*, (personal communication) also tested a long series of substances of the kind listed above with similar results. They did find, however, that temperature and humidity alone act as attractants, and they presented convincing evidence that a temperature-humidity differential between the host's body and the surrounding environment is essential to attraction.

Whatever the attractant may be, it is neither constant in the same individual at all times nor is it equally effective in all individuals. Experiments with *Anopheles gambiae* in West Africa (Ribbands, 1942) indicated that some human individuals are as much as five times more attractive than others inasmuch as they attracted five times as many mosquitoes under controlled conditions. There were also indications that considerable variation exists in the relative attractiveness of the same individual at different times. Another phenomenon noted by Ribbands was a summation of attractiveness. Under controlled conditions three men attracted from 2.2 to 2.8 times as many *A. gambiae* as one man and 2.5 times as many *A. funestus*.

· Little work has been done on the factors which assist predatory insects in locating their prey. While vision is undoubtedly the most widely employed sense here, those predators which are more or less particular in their selection of prey probably are assisted in locating it by odor. According to Murr (1930), *Habrobracon juglandis* thus locates its prey as does also *Philanthus triangulatus* (Tinbergen, 1932). Wesson's (1936) description of the manner in which the small predatory ant, *Strumigenys pergandei* stalks its prey is illustrative. This small ant scents a springtail (*Collembola*) 1 to 4 mm. away, stops, probes with its antennae, then slowly crawls till about 1 mm. away. Folding its antennae it then lowers its head to the ground and moves imperceptibly in the direction of the springtail until almost touching it with its mandibles. When the victim moves against the mandibles, the ant strikes and kills it.

On the whole, there is a nonspecificity of feeding habits among saprophagous insects, hence a similarity of ovipositional habits contrasting markedly with the precise specificity characteristic of other feeding groups such as the Phytophaga. Intergradations between the different categories, notably necrophagous and entomophagous species (see p. 28) are numerous. Considerable overlapping of spheres of attractiveness due to indiscriminate occurrence of many of the products under discussion may well explain intergradations of this sort. In the parts which follow the above-mentioned points are apparent.

Skatole, Indole, Mercaptans, Sulfides. Of the more nauseous odors of putrefaction those arising from skatole, indole, the mercaptans, and the sulfides are perhaps the most penetrating. Indole is a so-called heterocyclic compound of the formula

Skatole is its methyl derivative. Mercaptans are the sulfur analogs of alcohols, methyl mercaptan, for example, having the formula CH_3SH. Skatole has been tested as an attractant for many insects. Sexton beetles, *Necrophorus orbicollis* and *N. americanus*, and *Creophilus villosus* orient to and will dig up objects soaked in skatole (Abbott, 1936). Dung beetles of the genus *Geotrupes* react positively to skatole, indole, and ammonia (Warnke, 1931). Cheese mites, *Tyrolychus casei*, orient to skatole (Henschel, 1929). Many Diptera follow the odor of skatole to oviposition sites. Howlett (1912) induced oviposition by species of *Sarcophaga* with skatole. It is thus apparent that the prime attractive components directing necrophagous and especially coprophagous species to their food are skatole and indole. The possible effects of ammonia and the mercaptans deserve investigation, the more so since preliminary experiments (Dethier, unpublished) indicate that the former is not a powerful attractant of large African dung beetles.

Rather more information is available regarding substances attractive to blowflies and related Diptera. As pointed out earlier, it is difficult to distinguish between habitual and accidental entomiasis so that many transitions probably exist between carrion breeders and parasites. Perfectly fresh and slightly old animal tissues are not especially attractive to blowflies. There is a period during which meat baits are most attractive, and the time range of attractiveness varies with the species of insect. Parman, Laake, Bishopp, and Roark (1928) found, for example, that *Cochliomyia americana* prefers liver one day old and slightly infested with larvae, that species of *Lucilia* prefer fresh liver, that the range of positive response of *Lucilia* to liver in various stages of decomposition is broader than for *C. americana*. The fact that *Lucilia* is primarily a carrion breeder only secondarily parasitic, while *Cochliomyia* oviposits mostly in wounds and natural orifices, may in part explain the restricted range of positive response of the latter.

It was at one time assumed that unaltered sheep urine, dung, wax, suint, struck wool, and other animal products of this type played a large part in attracting sheep blowflies. Such does not prove to be the case. As Evans (1936) and Cragg and Ramage (1945) showed, attraction and stimulus to oviposition comes from products of bacterial activity upon these substrata. Evans also found that while carrion attracts both sexes, larval excreta attract females only, but that excreta must be placed on *living sheep* to incite oviposition (cf. p. 174). According to Hobson (1937), only gravid females are attracted to sheep treated with indole or ammonium carbonate. Cragg and Ramage found also that only females came to sheep wool plus ammonium carbonate, indole, and H_2S. Apparently any source of ammoniacal decomposition may constitute one factor inciting oviposition. *Lucilia cuprina* may also be stimulated to oviposit by the odor of flowers of *Stapelia flavirostris* (Hepburn and Nolte, 1943). At any rate, here are indications that two categories of attractants are operating, the one, attracting both sexes, the other, ovipositing females only.

Freney (1937) attempted to isolate highly attractive substances from carrion baits but without success. Many substances, however, have been found to be attractive to *Lucilia sericata* Meig., namely, low concentrations of methylallyl thiocyanate, amyl mercaptan (Hoskins and Craig, 1934), skatole, indole, ammonium carbonate (Hobson, 1936), and ethyl mercaptan (Freney, 1937). *Lucilia cuprina* is attracted by many of these substances and by NaOH and KOH (Hepburn and Nolte, 1943). In short, sheep blowflies are attracted by (1) the products formed by the hydrolysis with sodium sulfide of keratin, egg albumin, lecithin, and butter, namely, fatty acids, soaps, and sulfides; (2) any source of ammoniacal decomposition; (3) the mercaptans, indole, and skatole; (4) a few fatty acids such as butyric and valeric; (5) trimethylamine and isobutylamine; (6) certain organic sulfides (allyl sulfide, ethyl sulfide, ethyl disulfide), hydrogen sulfide, inorganic sulfides, and organic substances to which has been added Na_2S, $CaCO_3$, etc. It is of interest that the following substances *do not* attract sheep blowflies: acetic, propionic, isobutyric, caproic, caprylic, phenyl propionic, or lactic acids. Neither are dimethylamine, diethylamine, *n*-propylamine, *n*-amylamine, *n*-butylamine, piperidine, diphenylamine, nor allyl isothiocyanate attractive (Freney, 1937). Increased effectiveness of baits occasioned by the addition of sodium and alkyl sulfides or calcium carbonate (Fuller, 1934; Freney, 1937; Hepburn and Nolte, 1943) has not yet been satisfactorily or fully explained. Parman, *et al.*, (1928) noted, however, that calcium sulfide imparted an odor of H_2S to meat and this, being a normal attractant, drew flies. Freney showed experimentally that such is the case. Cragg and Ramage (1945) presented evidence that sulfur compounds, especially H_2S, are among the compounds normally produced by the bacterial decomposition of cystine, a major constituent of wool. The reaction may proceed as outlined below.

$$CH_2\!\!-\!\!S\!\!-\!\!S\!\!-\!\!CH_2 \longrightarrow CH_2\!\!-\!\!S\!\!-\!\!S\!\!-\!\!CH_2 \qquad CH_2SH$$
$$H_2N\ CHCO_2H \quad HO_2CCHNH_2 \quad CH_2R \qquad\quad CH_2R \longrightarrow CH_2R$$

Cystine · Disulfide · · · · · · Mercaptan

$$CH_3S\!\!-\!\!SCH_3 \longrightarrow CH_3SH$$

$$R = H,\ OH,\ NH_2,\ or\ CO_2H$$

Amines. As already noted, proteins may be hydrolyzed to amino acids, colorless crystalline compounds usually easily soluble in water. From amino acids are formed by decarboxylation common malodorous amines. Primary amines may be exemplified by methylamine, CH_3NH_2; secondary amines, by dimethylamine, $(CH_3)_2NH$; tertiary amines by trimethylamine, $(CH_3)_3N$. Other cosmopolitan putrefaction products of this nature are diamines of which putrescine, $NH_2CH_2CH_2CH_2CH_2NH_2$, and cadaverine, $NH_2CH_2CH_2CH_2CH_2CH_2NH_2$, are the better known. In the animal kingdom amines of all sorts are widespread; in the plant kingdom they are of less common occurrence. Outstanding examples of the presence of amines in green plants are the cotton plant (see p. 64), the flowers of hawthorn, the leaves of *Chenopodium vulvaria* and Dog's Mercury. Fungi, on the other hand, are replete with amines (Table 36). Unfortunately, a minimum amount of investigation has centered around the amines as attractants. An exception is Folsom's work with the Mexican boll weevil to which we have already alluded. As sex attractants amines apparently play an important rôle. Allylamine, ethylamine, and methylamine attract male gypsy moths (see p. 25); isoamylamine elicits sexual responses from male June beetles. The observations of Roubaud and Veillon (1922) that trimethylamine is attractive to some flies indicate that it may be of some importance as an attractant for coprophagous and necrophagous insects. Generally speaking, amines are unimportant as attractants for blowflies. There is an isolated report of the attractiveness of cyclohexylamine to *Lathyrophthalmus arvorum* Fabr. (Starr and Lau, 1938), but the significance of this compound is not apparent.

Fatty Acids. Probably better than any other compounds, fatty acids illustrate the manner in which spheres of attractiveness may overlap. These acids appear both as fermentation products and components of decomposing protein. In nature they are to be found in animal secretions and excretions, in living and decomposing green plants, in living fungi; they are one of the components of sex attractants; they attract certain ectoparasites of vertebrates (e.g., ticks orient to butyric acid); they induce oviposition in certain blood-sucking and saprophagous Diptera; they act as attractants for some phytophagous insects.

ACETIC ACID. In low concentrations acetic acid attracts some Diptera, especially species of *Drosophila*. As one of the common products of fermentation it attracts the codling moth and oriental fruit moth. For the most part, however, it acts as a repellent. When alcoholic fermentation attains the acetic phase, attractiveness is superseded by repellence.

PROPIONIC ACID. Few experiments have been conducted with this compound. It does not attract blowflies.

BUTYRIC ACID. This is one of the fatty acids, common in sweat and barnyard manure, serving as a sex attractant (to adult wireworms), as a feeding- and oviposition-type attractant (to many Calliphoridae, Sarcophagidae, Muscidae, Anthomyiidae, and such ectoparasites as ticks), and as a mild attractant to the codling moth, a fruit-feeder.

CAPROIC ACID. This acid was found to be attractive to adult male wire worms, probably in the guise of a sex attractant. It is exceptionally attractive to both sexes of the green June beetle, *Cotinus nitida* L. (Muma, 1944).

CAPRYLIC AND CAPRIC ACIDS. No experiments have been reported except those of Freney which indicated that caprylic acid does not attract sheep blowflies.

LAURIC AND MYRISTIC ACIDS. Studies of the reactions of copra beetles to copra (Corbett, Yusope, and Hassan, 1935) revealed that *Necrobia rufipes* De Geer, *Carpophilus dimidiatus*, and *Silvanus advena* are attracted more to slimy and moldy copra than to dry, and to decomposition products thereof. Of the fatty acid components of copra, lauric and myristic acids are especially attractive, more so than any others of the series. Mixed fatty acids are more attractive than coconut or palm oils *per se*.

LACTIC ACID. Because of its common occurrence and physiological significance, lactic acid has received much attention. In only a few cases has it proved of value as an attractant. As one of the products of the bacterial fermentation of milk (others include propionic, acetic, and carbonic acids) it serves well in this capacity with blowflies, chiefly the blue-bottle blowfly (*Calliphora erythrocephala* Meig.), the green-bottle blowfly (*Lucilia sericata* Meig.), and the black blowfly (*Phormia regina* Meig.) (McIndoo, 1933). It may also serve as a sex attractant for adult male wireworms.

VALERIC ACID. Four isomers, two of which occur in valerian root, are comprehended under this term. It has been found that the common one, present in barnyard· manure, is attractive to species of Muscidae, Calliphoridae, Sarcophagidae, and Anthomyiidae. It attracts adult male wireworms. It induces oviposition by *Stomoxys calcitrans* (Howlett, 1912).

UNSATURATED FATTY ACIDS. Of the unsaturated fatty acids, oleic and linolic have proved to be very good attractants for the oriental fruit moth (cf. Table 25) and for the copra beetle, *Silvanus advena*.

The true story behind the action of unsaturated fatty acids is obscured by their propensity for oxidizing spontaneously in the air. Mild oxidation results in the production of hydroxy acids as illustrated below:

    oleic   →dihydroxy stearic
    linolic   →tetrahydroxy stearic`
    linolenic→hexahydroxy stearic

Tests with these hydroxy acids might yield noteworthy results.

The attractiveness of fatty acids to certain other insects is indicated though not yet clearly understood. An accidental discovery by McPhail (1943) showed that fatty acids attract the melon fly. This species congregates on blossoms of a common orchid, *Dendrobium superbum*. Males are easily trapped when orchids or orchid extracts are used as bait. When an alcoholic extract attracted more females, it was discovered that the insects were actually attracted to the linseed oil soap solution used beneath the suspended test substance to soften the water and more firmly entrap the flies. Many soaps were then tested. Since linseed oil soap smells like freshly cut cucumbers, soap made from another host, seed of butter melon, was tried. It was not attractive. Linseed oil soap is a mixture of glycerides of fatty acids of which 0.5 to 1.5 per cent are unsaponifiable and 8 to 9 per cent are saturated acids. According to Holde (1922), the oil has the following composition: myristic, palmitic, stearic, and arachidic acids; 15 to 20 per cent oleic, 25 to 35 per cent linoleic, 35 to 45 per cent linolenic, 4 to 5 per cent glycine. On the strength of the fact that soaps made from the above acids (extracted from the oil) were nearly as attractive as linseed oil soap, it was concluded that the chief attractant is probably a fatty acid, but the fact that soaps made with chemically pure fatty acids are unattractive suggests the influence of some impurity.

Ammonia. Probably the most common and widespread single nitrogenous product of protein decomposition is ammonia. It occurs in feces (Table 30); it is a major constituent of urine where it forms 25 per cent of total urinary nitrogenous secretion; it is found in decaying vegetable matter and soil. In urine it is present not only as a primary product of excretion but also as a secondary product formed by the decomposition of urea. It is not surprising, then, that its importance as an attractant of saprophagous insects has long been suspected (see also under *Skatole, Indole, Mercaptans, Sulfides*). Its preëminence in this rôle over its two companion compounds, carbon dioxide and ethyl alcohol, has, however, been a bone of contention among different workers. As far as coprophagous insects are concerned, it has been fairly well established that ethyl alcohol is not an important agent. Alone it fails to attract species of *Rhyphus, Sarcophaga, Calliphora,* Muscidae, and Anthomyiidae (Imms and Husain, 1920). Both ethyl alcohol and acetic acid (4 to 10 per cent) are much less attractive than 4 per cent amyl alcohol (Richardson, 1917). McIndoo (1933) found acetic acid and carbon dioxide always repellent and alcohol up to 6 per cent only slightly attractive. According to Wieting and Hoskins (1939) ethyl alcohol is a feeble attractant for houseflies at a concentration of 0.012 per cent. In brief, freshly prepared carbohydrates and the simple fermentation products thereof are not especially powerful attractants for saprophagous insects feeding or ovipositing upon feces or carrion.

The relative importance of carbon dioxide *vs.* ammonia was in dispute for a long period, Richardson (1916, 1916a, 1916b, 1917, 1922) emphasizing the importance of the latter, Crumb and Lyon (1917, 1921), the former. Apparent discrepancies between the two schools of thought were finally dissipated by the precise experiments of Wieting and Hoskins (1939). Actually, both substances are attractive to a degree though ammonia is the more powerful. Divergent conclusions reached by different workers were due primarily to failure to consider the varying concentrations, as a gas, of the substances being tested. As has been frequently demonstrated (cf. Chapter 6), the effect of a given compound varies with its concentration. Depending on the concentration, the same compound may act as a repellent as well as an attractant. By means of a refined olfactometer designed to measure accurately the concentration of test gases, Wieting and Hoskins were able to make precise determinations on the attractiveness of ammonia, carbon dioxide, and ethyl alcohol to houseflies. Of the three substances, ammonia was the most attractive, the optimum concentration being 0.0120 per cent by volume in an air stream. Carbon dioxide was *almost* entirely unattractive at any concentration; ethyl alcohol only mildly attractive at 0.0126 per cent. Thus ammonia, whether it arises from manure in the field or is generated from ammonium carbonate or ammonium hydroxide in the laboratory, is the chief attractant inducing oviposition by the housefly. Attempts have been made in Russia to control flies by increasing the attractiveness of manure for oviposition through the addition of ammonium carbonate (Vanskaya, 1941). Presumably this is an adaptation of the trap crop technique. In field tests in this country it has been found that other insects, notably species of *Leptocera*, *Phorbia*, and *Muscina*, likewise assemble at baits of ammonium carbonate and other salts. Important discoveries have been made also concerning the effects of ammonia on species of fruit flies.

Many series of investigations directed toward the development of a commercial bait for fruit flies have produced valuable information pertaining to the attractants operating in the environment of these Diptera. Jarvis (1931) found that a combination of imitation essence of vanilla, ammonia, and water attracts both sexes of the Queensland fruit fly (*Chaetodacus tryoni*), the Jarvis fruit fly (*C. jarvisi*), the small black fly (*Dacus niger*), the solanum fruit fly (*C. dorsalis*), the boatman fruit fly (*Rioxa musae*), and the Mediterranean fruit fly (*Ceratitis capitata*). The success of Perkins and Hines (1934) in attracting *C. tryoni* with ammonia generated from ammonium carbonate, indicates that in all probability ammonia and not the vanilla essence was the important principle in the former case. Bua, in a long series of experiments (1933 and 1938) with the olive fruit fly, *Dacus oleae* Gmel., tested as attractants ammonium salts, several types of commercial molasses preparations called Dachicida and Dacivoro, combinations of these and ammonium salts, Clensel (an ammonia soap), cinnamon oil and borax, and an infusion of bran, dried figs, and borax. In the fall, the best results were obtained with a 10 per cent dilution of Dachicida P (molasses, 2.5 per cent sodium arsenite, 3.5 per cent water, 1 per cent olive oil, 0.5 per cent ammonium sulfate) and secondly with Clensel. In the late spring and summer a 3 per cent dilution of ammonium nitrate or phosphate, $(NH_4)_2HPO_4$, was most attractive. Bohorquez's (1940) results with the attractiveness of ammonium phosphate and fluoride to this insect are in agreement. In his traps females predominated.

With McPhail's (1939) discovery that fruit flies are attracted by mixtures of protein stuffs and sodium hydroxide, a more complete understanding of the chemotactic behavior of these Diptera became possible. Interest in protein lures originated with experiments designed to test the attractiveness of impure sugars undergoing hydrolysis. In Mexican brown sugar, the most attractive of those tested, the presence of protein was demonstrated when (1) the odor of ammonia followed heating in sodium hydroxide, (2) a proteinlike foam arose on sugar cooked in lime water, (3) sugar plus water plus alcohol precipitated a protein which when concentrated was very attractive. Proteins could be hydrolyzed equally well by alkalis or enzymes. In short, Trypetidae react positively to decomposing proteins when the products are those of mild hydrolysis. The chief product is ammonia in low concentrations. McPhail discovered that five species of fruit flies, including the Central American fruit fly, *Anastrepha striata*, are attracted by casein, gelatin, yeast, blood, egg white, wheat shorts, glycine, *dl*-alanine, and *l*-cystine in the presence of sodium hydroxide. Strangely enough, the Mexican fruit fly, *A. ludens*, does not respond. On the strength of McPhail's discoveries, Boyce and Bartlett (1941), who were seeking lures to collect data from which to determine the optimum time of treatment against the walnut husk fly, *Rhagoletis juglandis* Cress., initiated a series of trials with protein materials. They tested cane sugar, molasses, and Diamalt, each with brewer's yeast, wheat shorts with borax, casein with sodium hydroxide, fresh steer blood with water, and Clensel with water. Of the group, sugar with brewer's yeast and casein with sodium hydroxide showed the most promise, in the order named. Later, glycine (amino acetic acid) at 2.5 per cent in water, proved to be superior to casein. A further refinement produced a lure containing 2 per cent glycine (tech.) plus 3 per cent sodium hydroxide (commercial). Dean (1941) extended the idea of decomposing proteins as lures to the apple maggot fly, *Rhagoletis pomonella*. Of those tested the best attractants were powdered egg albumin, a peptone preparation, and dried yeast powder containing alkali. Eighty-three per cent of the catch was females. Benjamin and Hodson (1942) found that 60 per cent of their catch was females. Newman and O'Conner (1931) found Clensel in water (1:30) to be very attractive to the Mediterranean fruit fly. Although such amino acids as tyrosine, cystine, *dl*-alanine, glycine, *d*-glutamic and *l*-aspartic acids are slightly attractive in dilute solution, evidence suggests that further decomposition into ammonia undoubtedly is responsible for a high degree of attractiveness. The odor of ammonia is almost always present. In spite of the fact that a dilute solution of ammonia is only mildly attractive to *A. striata*, *A. suspensa* Loew, *A. mombinpraeoptans* Sein, *C. capitata*, and *Dacus cucurbitae* Coq., the weight of evidence points to ammonia as the most important attractive constituent of decomposing protein and the prime attractant of these fruit flies. It has been shown (Hodson, 1943) that 1 per cent household ammonia plus ammonium sulfate excels as a lure for the apple maggot fly. It is superior to fresh glycine. Just how the attractiveness of ammonia can be correlated with the natural behavior of the flies is not clear.

As adults, trypetid flies feed on dew, plant sap, and the like while as larvae they bore into living fruits and stems. In the latter stage of development they may exhibit marked feeding preferences. The degree of host plant specificity varies considerably with the species of fly. Larvae of the walnut husk fly are restricted to members of the genus *Juglans*. One strain of the apple maggot fly breeds only in blueberry and related fruits while another strain is confined to apple and its relatives. Some species are less restricted, e.g., the melon fly, which oviposits on tomatoes, watermelons, gourds, cucumbers, pumpkins, squashes, cantaloupes, etc. Still others are polyphagous. It would seem that feeding-type attractants for adults would be those compounds normally present in dew and plant sap, i.e., products of fermentation. The experiments of Ripley and Hepburn (see p. 24) strengthen this supposition, and examples are not uncommon. Honey and water (1:20) or wine, brown sugar, and water (1:2:40) are attractive to females of *D. cucurbitae* (Yasiro, 1940). A 40 per cent solution of arbutus juice is attractive to *D. oleae* (Bua, 1935). Other species are attracted to lures containing orange juice (Hayward, 1941). The Mexican fruit fly, *A. ludens*, is somewhat attracted by alcohol and strongly attracted by essence of white wine, fermenting sugar, cultures of wild yeast, and cultures of certain bacteria. To this fruit fly the products of protein decomposition are signally unattractive. By way of contrast, the products of fermentation are not especially attractive to species of flies strongly attracted to ammonia.

With regard to ovipositional-type attractants it would appear logical to expect flies to respond to compounds such as essential oils present in fresh fruit and stems. Walnut husk into which *R. juglandis* oviposits possesses a very penetrating odor due to an essential oil and some response would be expected. None has been found. Baker, *et al.* (1944) examined many extracts of the host plants of *A. ludens*, but all proved to be unattractive.

So for the moment the rôle of ammonia in the lives of these flies remains a mystery.

Miscellaneous Gases. Exclusive of the more pungent gases, notably ammonia and hydrogen sulfide, the attractive powers of which are a matter of record, the so-called odorless gases of low molecular weight are suspected of possessing some virtue as directive agents. Carbon dioxide, hydrogen, oxygen, and methane have been studied frequently in this respect.

As one of the by-products of the decomposition of organic material, carbon dioxide is almost without attractive powers to the housefly (Wieting and Hoskins, 1939). According to the early work of McIndoo (1933), however, all concentrations of CO_2 tested were repellent to blowflies. As a component of fermenting sugars CO_2 was found (Eyer and Medler, 1940) to act as a mild attractant to the codling moth. Oxygen and hydrogen likewise proved mildly attractive. Evidence has been put forward by Rudolfs (1922 and 1930), Huffaker and Back (1943), and Headlee (1945) to support the contention that CO_2 is an attractant for mosquitoes. CO_2 in mosquito traps is supposed to attract females. Males are said to be repelled (Huffaker and Back). The only other attractive so-called odorless gas, methane, also is a mosquito-attractant. It is said to encourage oviposition by *Culex pipiens*.

If CO_2 is to act either as an attractant or as a repellent, it must exceed in concentration that naturally present in the atmosphere, to wit, a few hundredths to several tenths of 1 per cent. The same prerequisite also holds true of oxygen, hydrogen, and methane. Much more careful work is required before the action of these gases as attractants can be established without a shadow of doubt.

Attractants of Mycetophagous Insects. It is indeed remarkable that so little work has been directed toward a study of attractants present in fungi. Opportunities are outstanding, especially with respect to insects feeding upon the larger, more conspicuous, forms. From the few existing examples of insects feeding directly on microscopic fungi we learn nothing of the existence of attractants directing or otherwise influencing plant choice because by and

FIG. 19. A portion of the substratum from the fungus garden of a Liberian termitarium showing the bromatia which constitute the edible portions of the fungus. (Photograph by Frank White.)

large there is no opportunity for choice on the part of the insect.
The fungus gardens of leaf-cutter ants and some termites are cases
in point.

Tropical and subtropical species of these two orders construct
within their nests elaborate gardens in which more or less pure
cultures of fungi are maintained. In termitaria the fungi are grown
on excreta. Leaf-cutter ants prepare an organic substratum of leaf
tissue liberally fertilized with excreta. Continued fungal growth is
maintained through inoculation by defecation since remnants of
ingested mycelia retain viability after passage through the ali-
mentary canal. No spores are ever formed. Bromatia, balls resulting
from the coiling and intertwining of hyphae to form sclerotiumlike
structures, constitute the edible portions of the plant. With the
founding of new nests, the fungus is established through contamination by the queen. Termites always carry the mycelia in their
guts; ants, in a special structure known as the infrabuccal pouch.
As the result of such an uninterrupted cycle in "closed gardens"
where no opportunity for a choice of menu is presented, feeding-
type attractants are hardly necessary. The fungi are thought to be
species of Agarics and Ascomycetes, especially *Xylaria*.

Beetles of the family Scolytidae, notably the ambrosia beetles,
may also be cited as examples of insects feeding directly on fungi
in ostensibly pure culture. The beetles breed in the sapwood of
dying trees where as larvae they feed on spores of a certain species
of fungus long known as *Monilia candida*. Recent work has shown
that different fungi may be cultivated by different species or groups
of species and that many kinds of fungi and yeasts may be associated
with beetles in various degrees of symbiosis. Verrall (1943) has
described four new species of fungi associated with southern species
of ambrosia beetles as follows: *Endomyces bispora* with *Platypus compositus*, *Cephalosporium pallidum* with *Xyleborus affinis*, *C. luteum* with *X.
pecanis*, and *Monilia brunnea* with *Pterocyclon mali* and *P. fasciatum*.
Wright (1935) has described a new wood-staining fungus, *Trichosporium symbioticum*, associated symbiotically with *Scolytus ventralis*.
Other relationships involving the blue stain fungi *Ceratostomella
pseudotsugae*, *C. piceaperda*, *C. ips*, and *C. montium* and various species
of beetles are reported by Rumbold (1936 and 1941). Good accounts of the interrelationships of insects and fungi, symbiotic and

otherwise, may be found in the works of Wheeler (1928), Leach (1940), and Leach, Orr and Christensen (1937).

Insects frequently are attracted to, and restrict their feeding to, some normal secretion products of living fungi. Many fungi, especially the rusts, depend upon a process analogous to insect pollination to insure propagation of the species. Specialized adaptations include bright orange pycnia, saccharine secretions, and characteristic accompanying odors. Flies, the usual agents of "diploidization," are attracted by the odors and color. While feeding on sugary droplets into which pycniospores have been exuded they effect spermatization. Specific examples include a disease of sugar cane known as gumming, the causative agent of which is a bacterium, *Phytomonas vascularum* (Cobb) Bergey, *et al.* The copious exudate initiated attracts large numbers of flies. Barth (1929) reported the presence of many species of *Agrotis* and other Noctuidae on the flowering spikes of *Elymus arenarius* L. to which they were attracted by an exudation occasioned by an agglomeration of conidia of *Claviceps purpurea* (Fr.) Tul., the active cause of the disease ergot. The sugary droplets secreted by the spores possess a distinct carrionlike odor, the chemical composition of which is undetermined. Quite probably it consists of one or more higher fatty acids or trimethylamine. Craigie (1931) has described similar phenomena as they occur in *Puccinia helianthi* Schw. and *P. graminis* Pers. In like manner are conidia of *Coprinus lagopus* transported from one heterothallic mycelium to another.

Attractants could be very satisfactorily studied in the larger fungi. These forms play host to a motley crowd of insects including the fungus gnats (Mycetophilidae), some psychodid flies, flies of the families Ceroplatidae and Platypezidae, the fungus beetles *Cybocephalus* and Endomychidae, all of which may undergo their development within the host plant. The number of insects that may be attracted to fresh and decaying species solely for feeding purposes is legion. In every case they are attracted by odors. All fungi possess a characteristic odor when fresh and a pronounced disagreeable odor when decaying. Certain ones even when fresh are characterized by foul, fetid, evil-smelling odors or stenches. Chief among these are the Gasteromycetes (*Dictyophora, Phallus, Aseroë, Clathrus, Mutinus* (*Phallalis*), *Rhizopogon, Hysterangium, Protubera*), Agaricales (*Russula foetida*), and Ustilaginales (*Tilletia*

Tritici). Fabre (1916) recounts most interestingly the behavior of certain insects seeking fungi. The beetle *Bolboceras gallicus* Muls. is able to locate very accurately by smell the underground fungus *Hydnocystis arenaria* Tul. toward which it burrows and which it consumes on the spot. In similar manner *Anisotoma cinnamomea* Panz. seeks truffles.

The chemistry of the larger more conspicuous fungi has received a fair amount of attention. As compared to the bewildering array of compounds making up the higher plants those of the fungi fall into a few well-defined groups. Carbohydrates of a dextrose nature are widespread; dyestuffs and ferments, common; proteins, present in small amounts; fats, richest as fatty acids; and bases, especially trimethylamine, widespread. Fungus membranes are composed of chitin or fungin. Chlorophyll and starches are lacking. Zellner (1907) differentiated the following groups from a chemical standpoint.

1. Saprophytic forms living on humus, from *Amanita* to *Fuligo*. This group is best known chemically.

    water content—85–94%
    fungins—small percentage
    fatty material—1–8%
    ergosterine group
    sugars
    trimethylamine group (choline, muscarine, etc.), enzymes

2. Coprophytic forms. The chemistry of this group is not especially well known. Urea groups occur here.

3. Wood-living forms. Of these the saprophytes are characterized by tannins and urea; the parasites, by low water content and a corky cell structure.

4, 5. Plant and animal parasites. Practically nothing is known of the chemistry of these groups.

The following complete analyses of four representative fungi given by Zellner illustrate the compounds present from which odors attractive to insects must be derived.

Amanita muscaria

    Water ±87–90%

    Fatty acids (free or as salts)—propionic, oleic, and palmitic acids ⎫
    Fats, glycerides of butyric, oleic, and palmitic acids ⎭ 0.9%

Lecithin 0.067%
Choline
Muscarine 0.016%
Muscaridine
Trimethylamine
Ergosterine 0.02–0.03%
Fumaric acid
Malic acid
Leucine
Mannite 0.7%
Crystalline carbohydrates
Amorphous carbohydrates
Fungin
Amorphous N-bearing bodies of unknown composition
Xanthine
Peptonelike bodies
Proteins in aqueous solution; in alkaline solution
Toxin
Dyestuffs
Tannic acid (?)
Amanitol
Ethereal oil
Fat-splitting enzymes
Proteolytic enzymes
Inverting enzymes
Mannite-building enzymes
Mineral material

Polyporus officinalis Fr.

Water 10%
Fatty acids ($C_{14}H_{24}O_2$ and $C_{18}H_{34}O_3$)
A substance, $C_{10}H_{16}O$, alcoholic in nature
Cetyl alcohol
An ergosterine substance
Oxalic acid
$C_4H_6O_4$
Fumaric acid 0.13%
Malic, tartaric, and an unknown acid
Trimethylamine
Glucose

Amorphous carbohydrate 6.8%
Fungin 20–30%
Resin 50–70%
Protein in aqueous solution, 0.7%; in alkaline solution
Ash 0.6–1.1%

Claviceps purpurea Tul.

Fatty acids: acetic, butyric, palmitic, oleic ⎫
Oxalic acid as a glyceride   ⎪
Glycerin   ⎬ 30%
Ergosterine   ⎪
                                            ⎭
Lactic acid
Leucine
Mycose 1%
Mannite
Fungin
Sclerojodin
Sclererythrin
Fuscosclerotic acid
Scleroxanthin
Scleropristallin
Methylamine
Trimethylamine
Choline
Betaine
Ergotinine
Ergotoxine
Sclerotic acid
Secalinaminosulfo acid
Chrysotoxine
Clavin
Vernine
Protein 2%
Ash 3–9%

Fuligo septica

Water 4.8%
Fatty acids: propionic, butyric, caproic, stearic, oleic
Salts of formic and acetic acids 0.42%
Salts of palmitic, stearic, and oleic acids 5.33%
Paracholesterin 1.4%

Lecithin 0.20%
Calcium oxalate 0.10%
Asparagine and another amido body 1.00%
Glycogen 4.73%
Sugar 3.00%
Guanine, xanthine, sarkine 0.01%
Plastin 27.40%
Pepsin and myosin 1.00%
Vitellin 5%
Peptone and peptonoid 4%
Resin 1.0%
Terpene
Dyestuffs, glycerin 0.18%
Calcium carbonate 27.70%
K_2HPO_4 121%
Iron phosphate 0.07%
$MgNH_4PO_4$ 1.44%
$Ca_3(PO_4)_2$ 0.91%
NaCl 0.10%
Ammonium carbonate 0.10%
Unknown substance 5%

A glance reveals that the most important of these, from the point of view of attractants, must be the higher fatty acids and amines. With the exception of water, a few inert substances like fungin and resin and, occasionally, relatively nonodorous metallic salts, fatty acids constitute a large percentage of the whole. In these and in amines are to be sought the attractants. Not only are they present in considerable quantity, but they are also of widespread occurrence (Tables 34, 35, 36). Fungal odors, usually referred to as stinking, carrionlike, rancid, fetid, fishy, ammoniacal, are the types characteristic of the higher fatty acids and trimethylamine. With the rapid onset of decay, odors become more vile because more fatty acids are freed and amines are broken down to skatole, indole, putrescine, and other stench-producing compounds. It is to be expected, then, that trimethylamine, butyric acid, palmitic acid, oleic acid, indole, and skatole may attract many mycetophagous insects. Reactions to propionic, caproic, and valeric acids and propylamine would also provide a fertile field of experimentation. Because of the unusually widespread occurrence of these compounds among the fungi one

Table 34

OCCURRENCE OF FATTY ACIDS IN THE FUNGI AND MYXOMYCETES*

Acid	Frequency	Occurrence
Formic........	Uncommon	*Lactarius vellereus* Fr.
		Polysaccum pisocarpium Fr.
Acetic.........		*Lactarius vellereus* Fr.
		Polysaccum pisocarpium Fr.
		Cantharellus cibarius Fr.
		Boletus edulis Bull.
		B. viscidus
		Phallus impudicus L.
		Hydnum repandum L.
Propionic......	Rare	*Fuligo septica* (L.) Gmelin
		Amanita muscaria L.
Butyric........	Frequent	*Amanita muscaria* L.
		Lactarius vellereus Fr.
		L. piperatus Scop.
		Cantharellus cibarius Fr.
		Boletus edulis Bull.
		Polysaccum pisocarpium Fr.
		Fuligo
Caproic........	Uncommon	*Fuligo*
Palmitic.......	Most common of this group	*Amanita pantherina* D. C.
		Boletus luridus Schaeff.
		Psalliota
		Fuligo
Stearic........		*Lactarius vellereus* Fr.
Oleic..........	Very common in free state	*Amanita pantherina* D. C.
		Fuligo septica (L.) Gmelin
		Boletus luridus Schaeff.
		Lactarius vellereus Fr.

* Data from Zellner, 1907.

Table 35

OCCURRENCE OF OTHER ACIDS IN THE FUNGI AND MYXOMYCETES*

Acid	Frequency	Occurrence
Oxalic.......	Widespread, especially as salts	*Fuligo septica* (L.) Gmelin
Lactic.......	..	*Fuligo septica* (L.) Gmelin
Succinic......	Uncommon	..
Fumaric.....	Very widespread	..
Malic.......	Uncommon	..
Tartaric......	In one species only	*Cantharellus cibarius* Fr.
Citric........	Uncommon	*Amanita*

* Data from Zellner, 1907.

may assume that they act as rather general attractants and that mycetophagous insects (fungus-growing species excepted) are essentially polyphagous. If mycetophagous insects are truly attracted by some of these substances, there is reason to expect that they would congregate around the flowers of hawthorn which owe their unpleasant odor to trimethylamine and propylamine or around some species of *Aristolochia* the fetid odor of which derives from indole. The entire subject has been greatly neglected.

Table 36

OCCURRENCE OF AMINO ACIDS AND AMINES IN THE FUNGI
AND MYXOMYCETES*

Compound	Frequency	Occurrence
Leucine		*Amanita muscaria* L.
Asparagine	Uncommon	*Fuligo*
Glutamine		*Fuligo*
Tyrosine		*Lycoperdon Bovista* L.
Urea		*Lycoperdon Bovista* L.
Methylamine	Common	*Polyporus officinalis*
		Claviceps
Trimethylamine	Very widespread	
Choline		Higher fungi
Muscarine		Higher fungi

* Data from Zellner, 1907.

Attractants of Phycophagous Insects. The aquatic habitat of algae greatly limits the number of insect species that derive sustenance from these plants. With the exception of a few, such as the seaside fly, *Scatella subquattata*, which as an adult eats green algae (Brauns, 1939), most of the insects preying upon living algae do so in the larval stage. Of these, dipterous larvae constitute the greatest percentage. Numbered among the more important forms are larvae of mosquitoes, simuliids, the psychodids *Phlebotomus papataci* and *P. minutus* which in Asia Minor feed on half-dried algae in cisterns (Brunetti, 1913), the ceratopogonine midge *Stilobezzia spirogyrae* which is associated with *Spirogyra* (Carter, Ingram, and Macfie, 1921), the Dixidae, and the marine chironomid *Pontomyia natans* which feeds on diatoms (Buxton, 1926). The last mentioned insect is one of the few associated with a living marine alga.

Our knowledge of the food preferences and attractants operating in the choice of food is scanty to the point of nonexistence even for economically important species. The question of whether or not mosquito larvae exercise choice in the selection of food has excited voluminous controversy. Many authors argue in the affirmative. Lamborn (1921), for example, states that *Anopheles maculatus* and *A. karwari* have a predilection for certain filamentous algae, particularly a species of *Spirogyra*. Alessandrini (1925) maintained that anopheline larvae in rice fields have a decided preference for algae rich in chlorophyll and that a gross selection of algal species rich in chlorophyll might be made on the basis of color. Most of the evidence, however, points to the absence of preferential feeding. Rudolfs (1925), Senior-White (1928), Hinman (1930) and others have shown experimentally and by comparing gut contents with plants available in the water in which larvae were living that (1) food supply is determined by the chemical composition of the water, (2) larvae will eat whatever is available, and (3) the mechanical manner of feeding, namely creation of a current which sweeps everything into the mouth, does not make for selective feeding. Thus, ideas of preference based on constant association of larvae with certain algal species must be accepted with reservations.

Certain species of green algae, notably *Chara foetida, C. contraria, C. hispida, C. fragilis,* and *C. intermedia,* are said to be repellent to larvae. Matheson and Hinman (1930) postulated that larvae were killed by ingesting oxygen bubbles generated by *Chara* and in 1931 showed that larvae can be killed by passing small bubbles of air through water. It has been shown also that *Chara* grown in the dark is not detrimental to larvae. These experiments would indicate that any repellence is of a physical nature. The possibility of a chemical repellent has not been investigated at any length. The chemical properties of some of the algae that impart odors to drinking water should be examined and the reactions of mosquito larvae to certain of these studied. The possible presence of chemical attractants likewise has not been investigated though it is known that some diatomaceae possess an aromatic or geranium odor while others possess a fishy odor. Also, certain Chlorophyceae have a fishy odor, while the Cyanophyceae emanate a grassy odor.

Discussion. With the accumulation of facts concerning the interrelations of saprogenic attractants and insects there have arisen new questions upon the solution of which depends a clearer understanding of this subject as a whole. Five of the more important ones may be stated. (1) What attractive principles guide mosquitoes to their hosts? (2) Is carbon dioxide an insect-attractant? (3) In what manner do temperature and humidity affect the behavior of insects toward their hosts? (4) What are the factors attracting blowflies to their hosts? (5) In what way is the action of ammonia correlated with the normal behavior of fruit flies?

1. and 2. Parasitic, blood-sucking, and predatory insects usually are attracted to their hosts or prey by visual factors, temperature, or odors acting independently or in combination. Early assumptions that mosquitoes too were profoundly influenced by host odors have not withstood continued investigation. Neither the body tissues of hosts, secretions, excretions, nor their individual constituents exhibit any marked degree of attractiveness for mosquitoes. Blood is signally ineffective. One metabolite, however, is the subject of extensive controversy. This is carbon dioxide. Rudolfs, Headlee, and Huffaker and Back on the one hand claim that carbon dioxide is capable of attracting mosquitoes under conditions of optimum temperature and humidity. Crumb (1922), and Davidson, Venard, Peffly and DeLong (personal communication) on the other hand found that carbon dioxide exerts no powers of attraction. Wieting and Hoskins, working with houseflies in an olfactometer, stated that carbon dioxide is almost entirely unattractive. Eyer and Medler, experimenting with codling moths in an olfactometer, stated that carbon dioxide is more than a neutral carrier and possibly even a mild attractant. Without further careful experimentation this controversy cannot be decided, but the following pertinent facts can be pointed out: (1) in field experiments which showed carbon dioxide to be attractive the gas was either combined with other chemicals or was dispensed under special conditions of temperature and humidity; (2) in olfactometer tests under controlled conditions carbon dioxide was at most a questionable attractant and may have been acting as a stimulus to over-all activity; (3) to be an effective attractant (or repellent) the concentration of carbon dioxide must exceed that normally present in the atmosphere yet must not be so great as to anesthetize.

Strong evidence that a temperature-humidity factor and not carbon dioxide is the principal attractant directing mosquitoes to their hosts has been offered by Crumb and by Davidson, *et al.* The latter have shown that a temperature-humidity differential between the host's body and the surrounding environment is necessary for attraction and that no attraction exists in the absence of such a differential. By comparing two dishes of water at 98.6° F., one sealed with a Baudruche membrane permeable to water and the other sealed with an impermeable pure latex cap, they demonstrated the action of combined temperature and humidity as an attractant.

3. It is common knowledge that insects will place themselves in zones of preferred humidity and that they react to small differences of RH in the preferred range. Furthermore, it appears that with some species the reaction is initiated by evaporation and does not involve hygrometer receptors. This raises the question of whether or not humidity reactions are sufficiently rapid to permit of orientation to a host and whether a tactic type of response is possible. In the case of sheep blowflies high humidity is essential to the complete success of many attractants, but the manner in which it acts is not known. Careful tests have shown that wool to which have been added attractants known to induce oviposition on live sheep (ammonium carbonate and indole) fails to incite oviposition unless wet. From these experiments and from those with mosquitoes it does seem that humidity deserves further consideration as an attractant.

4. The precise nature of ovipositional-type attractants and of stimuli affecting blowflies is almost as controversial a subject as those just discussed. Most workers are now agreed that unaltered animal tissues and products such as dung, wax, wool, and suint are unattractive but that they become attractive after various degrees of decomposition. It is agreed also that the addition of sulfur compounds to baits increases their attractiveness. Disagreements arise chiefly over the nature of the stimuli necessary to induce oviposition and over the attractiveness of mercaptans.

The postulation of some specific factor associated with living sheep as essential for attraction and oviposition is not absolutely necessary. Loeffler and Hoskins (1943), and Cragg and Ramage (1945), have shown that attraction may be partially independent of any factor of this sort. Evans' statement that larval excreta must be placed on living sheep to induce oviposition may find explanation in the conclusions of Cragg and Ramage that a *combination* of ammonia and sulfur compounds is necessary to oviposition and that larval excreta contain chiefly H_2S. A solution of 0.5 per cent ammonium carbonate plus 0.01 per cent ethyl mercaptan or 0.1 per cent ammonium carbonate plus 0.002 per cent ethyl mercaptan or 0.5 per cent ammonium carbonate plus 0.1 per cent indole will induce oviposition by *Lucilia sericata* on living sheep. Either constituent alone is ineffective. Ammonium carbonate plus indole will induce oviposition on moist clipped fleece.

From these data it seems clear that sheep wool (but not living sheep) is necessary for oviposition, that a combination of ammonia and sulfur compounds is necessary to produce an ovipositional-type attractant, and that only females are attracted. Data presented elsewhere indicated that carrion attracts both sexes. From this it is apparent that sheep blowflies are subjected to at least two distinct and different categories of attractants, food-type and ovipositional-type.

Disagreements involving mercaptans as attractants for blowflies have not yet been resolved. In general, the sulfur compounds have not been subjected to extensive study. The only detailed cases exclusive of that just described are those involving interrelations between mustard oils and *Pieris* larvae, alkyl sulfides and *Allium*-feeders, and mercaptans and coprophagous species. An unusual case is that reported by Baker, *et al.* (1944) in which Mexican fruit flies were especially strongly attracted to a culture of bacteria obtained from uninjured mango, a culture that was characterized by a slight odor of H_2S.

5. Ammonia, of all the compounds arising from the decomposition of fats and proteins, appears as the most important single attractant in nature. It is attractive to some degree to blood-sucking, saprophagous, coprophagous, necrophagous, mycetophagous, and carpophagous insects. It is one compound the action of which is usually clearly understood. Its widespread attraction may be attributed in part to its ubiquitous occurrence. As a component of cotton it attracts boll weevils, as a characteristic compound in manure it attracts various species of flies. The one association which is not understood is that of ammonia and fruit flies. No satisfactory explanation is apparent.

Included among other compounds containing nitrogen are skatole, indole, and the amines. The former as products of advanced putrefaction attract chiefly necrophagous and coprophagous insects whereas ammonia is active with insects of the saprophagous category. It has been suggested that indole and skatole may act by "fixing" odors rather than as true attractants. This idea is worthy of further investigation. Amines, too, must be classed with those compounds deserving intensive investigation. They and the fatty acids are undoubtedly of great importance as regards the behavior of mycetophagous insects.

REFERENCES

Abbott, C. E.: On the olfactory powers of a necrophilous beetle, *Bull. Brooklyn Entomol. Soc.*, **31**(2), 73–75 (1936).

Alessandrini, G.: Lo Studio biologico della Risaia. *Ann. igiene*, **35**(10), 912–914 (1925); *Rev. Applied Entomol., B* **14**, 15–16 (1925).

Baker, A. C., W. E. Stone, C. C. Plummer, and M. McPhail: A review of studies on the Mexican fruitfly and related Mexican species, *U.S. Dep. Agr., Misc. Pub.*, 531 (1944).

Bard, P.: "MacLeod's Physiology in Modern Medicine," St. Louis, C. V. Mosby Co., 1938.

Barth, G.: Eulenfang am Honigtau (Lep.), *Entomol. Z., Frankfurt a/M*, **43**(18), 224–225 (1929).

Benjamin, D. M., and A. C. Hodson: The apple-maggot problem again, *Minnesota Hort.*, **70**(3), 54–55 (1942).

Bohorquez, R.: Experiencias de lucha contra la mosca del olivo (*Dacus oleae* Gmel.) por medio de sustancias atractivas, *Bol. pat. veg. ent. agr.*, **9**, 188–204 (1940).

Boyce, A. M.: The walnut husk fly (*Rhagoletis juglandis* Cresson), *J. Econ. Entomology*, **22**(6), 861–866 (1929).

——, and B. R. Bartlett: Lures for the walnut husk fly, *Ibid.*, **34**(2), 318 (1941).

Bradley, G. H.: Some factors associated with the breeding of Anopheles mosquitoes, *J. Agr. Research*, **44**(5), 381–399 (1932).

Brauns, A.: Zur Biologie der Meeresstrandfliege *Scatella subguttata* Meig. (Familie Ephydridae; Diptera), *Zool. Anz.*, **126**(11/12), 273–285 (1939).

Brighenti, D.: Sull'alimentazione della larve di *Anopheles*, *Bol. zool.*, **2**(2), 27–31 (1931).

Browne, C. A.: The chemical analysis of the apple and some of its products, *Penn. State College, Ann. Rep.*, **1899–1900**, 262–276, 1901.

*Brunetti, C.: Some noxious Diptera from Galilee, *J. Proc. Asiatic Soc. Bengal*, **9**(1), 43–45 (1913).

Bua, G.: Experimenti dell 1933 con sostanze attrative per la mosca delle olive, *Ann. ist. super. agrar. Portici*, **6**(3), 125–145 (1933).

——: Seconda serie di esperimenti con sostanze attrative per la mosca delle frutta, "Ceratitis capitata" Wied., *Bol. Lab. zool. gen. agrar. Portici*, **28**, 295–308 (1935).

——: Serie di esperimenti con sostanze attrative per la mosca delle olive, *L'Olivicoltore*, **15**, 1–19 (1938).

Buxton, P. A.: The colonization of the sea by insects, *Proc. Zool. Soc. London*, **1926**, 807–814.

Carter, H. F., A. Ingram, and J. W. S. MacFie: Observations on the ceratopogonine midges of the Gold Coast, with descriptions of new species, III, *Ann. Trop. Med. Paras.*, **14**(3), 309–331 (1921).

Corbett, G. H., M. Yusope, and A. Hassan: The attraction of *Necrobia rufipes* de Geer (the copra beetle) to the fatty acid of coconut oil and to types of copra. *Malayan Agr. J.*, **23**(5), 217–228 (1935).

Cragg, J. B., and G. R. Ramage: Chemotropic studies on the blowflies *Lucilia sericata* (Mg.) and *Lucilia caesar* (L.), *Parasitology*, **36**(3–4), 168–175 (1945).

Craigie, J. H.: An experimental investigation of sex in the rust fungi, *Phytopathology*, **21**(11), 1001–1040 (1931).

Crombie, A. C.: On oviposition, olfactory conditioning and host selection in *Rhizopertha dominica* Fab. (Insecta, Coleoptera), *J. Exp. Biol.*, **18**(1), 62–79 (1941).

Crumb, S. E.: A mosquito attractant, *Science*, **55**(1426), 446–447 (1922).

——, and S. C. Lyon: The effect of certain chemicals upon oviposition in the house-fly (*Musca domestica* L.), *J. Econ. Entomol.*, **10**(6), 532–536 (1917).

——, and ——: Further observations on the effect of certain chemicals upon oviposition in the housefly (*Musca domestica* L.), *Ibid.*, **14**(6), 461–465 (1921).

Davidson, R. H., C. E. Venard, R. L. Peffly, and D. M. DeLong: (Personal communication, 1945.)

Dean, R. W.: Attraction of *Rhagoletis pomonella* adults to protein baits, *Ibid.*, **34**(1), 123 (1941).

Evans, A. C.: The physiology of sheep blow-fly *Lucilia sericata* Meig. (Diptera), *Trans. Entomol. Soc. London*, **85**(15), 363–378 (1936).

Fabre, J. H.: "The Life of the Caterpillar," New York, Dodd, Mead, & Co., 1916.

Freney, M. R.: The chemical treatment of baits for attracting blowflies, *J. Council Sci. Ind. Research*, **5**, 94–97 (1932).

——: Studies on the merino fleece. I. The chemistry of suint, *J. Soc. Chem. Ind.*, **53**(18), 131–134 (1934).

——: Studies on the chemotropic behavior of sheep blowflies, *Australian Council Sci. Ind. Research*, *Pamphlet* 74, 1937.

Fuller, M. E.: Sheep blowfly investigations. Some field tests of baits treated with sodium sulfide, *J. Council Sci. Ind. Research*, **7**, 147–149 (1934).

Gortner, R. A.: "Outlines of Biochemistry," 2d ed., New York, John Wiley & Sons, 1938.

Graenicher, S.: Some biological notes on *Sarcophaga bullata* Park. (Diptera: Sarcophagidae), *Entomol. News*, **46**(7), 193–196 (1935).

Harrow, B., and C. P. Sherwin: "A Textbook of Biochemistry," Philadelphia, W. B. Saunders Co., 1935.

Hayes, T. H.: Report of mosquito survey in St. Croix, *U.S. Naval Med. Bull.*, **28**(1), 194–222 (1930).

Hayward, K. J.: La lucha contra las moscas de las frutas, *Rev. Ind. Agr. Tucuman*, **31**, 331–349 (1941).

Headlee, T. J.: "The Mosquitoes of New Jersey and Their Control," Rutgers University Press, 1945.

Henschel, J.: Reizphysiologische Untersuchungen an der Käsemilbe *Tyrolichus casei* (Oudemans). *Z. vergl. Physiol.*, **9**(5), 802–837 (1929).

Hepburn, G. A., and M. C. A. Nolte: Sheep blowfly research. III. Studies on the olfactory reactions of sheep blowflies, *Onderstepoort J. Vet. Sci. Animal Ind.*, **18**(1), 27–48 (1943).

——, and ——: Sheep blowfly research. IV. Field tests with chemically treated carcasses, *Ibid.*, **18**(1), 49–57 (1943a).

Heymons, R., and H. v. Lengerken: Biologische Untersuchungen an coprophagen Lamellicorniern, *Z. Morph. Ökol. Tiere*, **14**, 531–613 (1929).

Hinman, E. H.: A study of the food of mosquito larvae (Culicidae), *Am. J. Hyg.*, **12**(1), 238–270 (1930).

Hobson, R. P.: Sheep blow-fly investigations. III. Observations on the chemo-tropism of *Lucilia sericata* Mg., *Ann. Applied Biol.*, **23**(4), 845–851 (1936).

——: Sheep blow-fly investigations. IV. The chemistry of the fleece with reference to the susceptibility of sheep to blow-fly attack, *Ann. Applied Biol.*, **23**(4), 852–861 (1936a).

——: Sheep blow-fly investigations. V. Chemotropic tests carried out in 1936, *Ibid.*, **24**(3), 627–631 (1937).

——: Sheep blow-fly investigations. VIII. Observations on larvicides and repel-lents for protecting sheep from attack, *Ibid.*, **27**(4), 527–532 (1940).

——: II. Recent work on the sheep maggot problem, *Ibid.*, **28**(3), 297–299 (1941).

Hodson, A. C.: Lures attractive to the apple maggot, *J. Econ. Entomol.*, **36**(4), 545–548 (1943).

Holde, D.: "The Examination of Hydrocarbon Oils and of Saponifiable Fats and Waxes," 2d ed., London, Chapman and Hall, 1922.

Hoskins, W. M., and R. Craig: The olfactory responses of flies in a new type of insect olfactometer, *J. Econ. Entomol.*, **27**(5), 1029–1036 (1934).

Howlett, F. M.: The effect of oil of citronella on two species of *Dacus, Trans. Entomol. Soc. London*, **1912**, 412–418.

Huffaker, C. B., and R. C. Back: A study of methods of sampling mosquito populations, *J. Econ. Entomol.*, **36**(4), 561–569 (1943).

Hutner, S. H., H. M. Kaplan, and E. V. Enzmann: Chemicals attracting *Dro-sophila, Am. Naturalist*, **71**, 575–581 (1937).

Imms, A. D., and M. A. Husain: Field experiments on the chemotropic responses of insects, *Ann. Applied Biol.*, **6**(4), 269–292 (1920).

Jarvis, H.: Experiments with a new fruit fly lure, *Queensland Agr. J.*, **36**(5), 485–491 (1931).

Lamborn, W. A.: Federated Malay States Malaria Bureau Report, 1920; Suppl. to F. M. S. Gov.-Gaz., 1921, pp. 8–13, *Rev. Applied Entomol.*, B **10**, 57–58 (1921).

Leach, J. G.: "Insect Transmission of Plant Diseases," New York, McGraw-Hill Co., 1940.

——, L. W. Orr, and C. Christensen: Further studies on the interrelationships of insects and fungi in the deterioration of felled Norway pine logs, *J. Agr. Research*, **55**(2), 129–140 (1937).

McIndoo, N. E.: Olfactory responses of blowflies, with and without antennae, in a wooden olfactometer, *Ibid.*, **46**(7), 607–625 (1933).

McPhail, M.: Protein lures for fruitflies, *J. Econ. Entomol.*, **32**(6), 758–761 (1939).

——: Linseed oil soap—A new lure for the melon fly, *Ibid.*, **36**(3), 426–429 (1943).

Marchand, W.: First account of thermotropism in *Anopheles punctipennis*, with bionomic observations, *Psyche*, **25**(6), 130–135 (1918).

Marquez, V. M.: Experiencias sobre el poder atrayente, para la mosca del olivo, del fos fato a diverses concentraciones, *Bol. Path. Veg. Ent. Agr.*, **10**, 237–242, *Rev. Applied Entomol.*, A, **30**, 480 (1941).

Matheson, R., and E. H. Hinman: A seasonal study of the plancton of a spring-fed *Chara* pool versus that of a temporary to semipermanent woodland pool in relation to mosquito breeding, *Am. J. Hyg.*, **11**(1), 174–188 (1930).

——, and——: Further work on *Chara* spp. and other biological notes on Culicidae (Mosquitoes), *Ibid.*, **14**(1), 99–108 (1931).

Mickelson, O., and A. Keys: The composition of sweat, with special reference to the vitamins, *J. Biol. Chem.*, **149**(2), 479–490 (1943).

Mönnig, H. O.: Sheep blowfly research. VI. The treatment of myiasis, *Onderstepoort J. Vet. Sci. Animal Ind.* **18**(1), 73–84 (1943).

Muma, M. H.: The attraction of *Cotinus nitida* by caproic acid, *J. Econ. Entomol.*, **37**(6), 855–856 (1944).

Murr, L.: Über den Geruchssinn der Mehlmottenschlupfwespe *Habrobracon juglandis* Ashmead, *Z. vergl. Physiol.*, **11**(2), 210–270 (1930).

Newman, L. J.: Fruit-fly (*C. capitata*): Trapping or luring experiments, *J. Dep. Agr. W. Australia*, Ser. 2, **3**(4), 513–515 (1926).

——, and B. A. O'Connor: Fruit fly (*Ceratitis capitata*). A further series of trapping or luring experiments, *Ibid.*, Ser. 2, **8**, 316–318 (1931).

Nolte, M. C. A.: Sheep blowfly research. II. Suint investigations, *Onderstepoort J. Vet. Sci. Animal Ind.* **18**(1), 19–25 (1943).

Parman, D. C., E. W. Laake, F. C. Bishopp, and R. C. Roark: Test of blowfly baits and repellents during 1926, *U.S. Dep. Agr., Tech. Bull.* 80, 1–14 (1928).

Perkins, F. A., and H. J. Hines: A note on some preliminary experiments with ammonia as a lure for the Queensland fruit fly (*Chaetodacus tryoni* Frogg), *Proc. Roy. Soc. Queensland*, **45**, 29 (1934).

Ribbands, C.: Progress Report—Experimental Entomology, No. 5 Malaria Field Laboratory, Roy. Army Med. Corps, July 28, 1942.

Richardson, C. H.: The attraction of diptera to ammonia, *J. Econ. Entomol*, **9**(4), 408–413 (1916).

——: A chemotropic response of the house-fly (*Musca domestica* L.), *Science*, **43**(1113), 613–616 (1916a).

——: The response of the house-fly (*Musca domestica* L.) to ammonia and other substances, *New Jersey Agr. Exp. Sta., Bull.* 292, 1–19 (1916b).

——: The response of the house-fly to certain foods and their fermentation products, *J. Econ. Entomol.*, **10**(1), 102–109 (1917).

——, and E. H. Richardson: Is the house-fly in its natural environment attracted to carbon dioxide? *Ibid.*, **15**(6), 425–430 (1922).

Roubaud, E., and R. Veillon: Recherches su l'attraction des mouches communes par les substances de fermentation et de putréfaction, *Ann. Inst. Pasteur*, **36**, 752–764 (1922).

Rudolfs, W.: Chemotropism of mosquitoes, *New Jersey Agr. Exp. Sta., Bull.* 367, 1–23 (1922).

——: The food of mosquito larvae, *Proc. 12th Annual Meeting New Jersey Mosq. Exterm. Assoc.*, **1925**, 25–30.

——: Effect of chemicals upon the behavior of mosquitoes. *New Jersey Agr. Exp. Sta., Bull.* 496, 1–24 (1930).

Rumbold, C. T.: Three blue-staining fungi, including two new species, associated with bark beetles, *J. Agr. Research*, **52**(6), 419–437 (1936).

——: The blue stain fungus, *Ceratostomella montium* N. Sp., and some yeasts associated with two species of *Dendroctonus*, *Ibid.*, **62**(10), 589–601 (1941).

Senior-White, R.: Algae and the food of Anopheline larvae, *Indian J. Med. Research*, **15**(4), 969–988 (1928).

Severin, H. H. P., and H. C. Severin: Relative attractiveness of vegetable, animal, and petroleum oils for the Mediterranean fruit fly (*Ceratitis capitata* Wied.), *J. New York Entomol. Soc.*, **22**(3), 240–248 (1914).

Speyer, E. R.: Notes on the chemotropism in the house-fly, *Ann. Applied Biol.*, **7**(1), 124–140 (1920).

Starr, D., and K. H. Lau: Cyclohexylamine, an attractant for *Lathyrophthalmus arvorum* (Fabr.), *Proc. Hawaiian Entomol. Soc.*, **10**(1), 61 (1938).

Tinbergen, N.: Über die Orientierung des Bienenwolfes (*Philanthus triangulum* Fabr.), *Z. vergl. Physiol.*, **16**(3), 305–334 (1932).

Vanderplank, F. L.: Studies of the behaviour of the tsetse-fly (*Glossina pallidipes*) in the field: the attractiveness of various baits, *J. Animal Ecol.*, **13**(1), 39–48 (1944).

Vanskaya, R. A.: The use of $(NH_4)_2CO_3$ for the control of *Musca domestica*, *Med. Parasitol. Parasitic Diseases* (*U.S.S.R.*), **10**, 562–567 (In Russian) (1941).

Verrall, A. F.: Fungi associated with certain ambrosia beetles, *J. Agr. Research*, **66**(3), 135–144 (1943).

Warnke, G.: Experimentelle Untersuchungen über den Geruchssinn von *Geotrupes sylvaticus* Panz. und *Geotrupes vernalis* Lin., *Z. vergl. Physiol.*, **14**(1), 121–199 (1931).

Weber, H.: Biologische Untersuchungen an der Schweinlaus (*Haematopinus suis* L.) unter besonderer Berücksichtigung der Sinnesphysiologie, *Ibid.*, **9**(4), 564–612 (1929).

Wesson, L. G.: Contributions toward the biology of *Strumigenys pergandei:* A new food relationship among ants, (Hymen: Formicidae), *Entomol. News*, **47**(7), 171–174 (1936).

Wheeler, W. M.: "The Social Insects," New York, Harcourt, Brace & Co., 1928.

Wieting, J. O. G., and W. M. Hoskins: The olfactory responses of flies in a new type of insect olfactometer, *J. Econ. Entomol.*, **32**(1), 24–29 (1939).

Wilson, H. F.: Lures and traps to control clothes moths and carpet beetles, *Ibid.*, **33**(4), 651–653 (1940).

Wright, E.: *Trichosporium symbioticum*, N. Sp., a wood-staining fungus associated with *Scolytus ventralis*, *J. Agr. Research*, **50**(6), 525–538 (1935).

Yasiro, H.: Baits for attracting *Dacus cucurbitae*. Ôyò-Kontyù, **2**, 162–165 (in Japanese) (1940).

Zellner, J.: "Chemie der höheren Pilze," W. Engelmann, Leipzig, 1907.

Chapter 6

Olfactometers and Threshold Concentrations

Measurement of Odors. The major part of our knowledge of attractants has been derived from observations and from experiments based upon one or more of three standard fundamental procedures. To ascertain whether or not a given substance is attractive or to compare and evaluate suspected or actual attractants one may resort to the use of (1) feeding tests, (2) bait traps, or (3) olfactometers. By the first method the substance to be tested, whether it be a natural food, treated food, or ersatz food, is offered directly to the insect. The response of the insect and the amount of food eaten is taken as an index of the attractance or repellence of the material in question. Various methods of preparing these tests and analyzing the results are in use. Some of the spraying and sandwich methods tried with phytophagous insects have already been described in Chapter 2. Since similar methods are in extensive use for testing repellents, additional discussion has been deferred till a later chapter. Results are largely comparative.

Bait traps offer an arrangement whereby test substances are exposed in the field to a normal insect population existing in its normal environment. Therein lie the advantages. Disadvantageous is the fact that results are entirely qualitative and the effective concentration of test materials cannot be accurately determined or controlled. The value, proper function, and use of bait traps is treated in Chapter 7.

To determine accurately and quantitatively the attractive or repellent properties of any material it is necessary to resort to the use of intricate apparatus in the laboratory. This is a refinement of the timeworn procedure of holding some odorous substance near an insect. Coincident with attempts to establish a science of odors have been equally persistent attempts to produce a machine by means of which an organism's responses to odors could be accurately catalogued and evaluated. Almost as numerous as the models invented have been the names ascribed to them. Among those proposed have

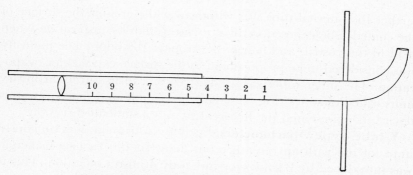

FIG. 20. Diagram of Zwaardemaker's olfactometer.

been chemotropometer, odorometer, and olfactometer. The selection of a fit name depends upon the avowed purpose and performance of the apparatus. The intentions behind the design of the different instruments have by no means been identical yet a single goal is being sought. What is desired is an instrument by means of which the nature of an organism's responses to odors of a known concentration may be determined. Since none of the names previously proposed are fit, we suggest that, rather than propose a new name, the name olfactometer be retained on account of convenience arising from its widespread common usage and early origin. The name as originally applied and now retained is not correct, however, since it measures neither the odors in the sense that a thermometer measures temperature, nor the act of smelling or the resultant response. The word chemotropometer is equally incorrect in that it, too, does not measure response to chemicals. The word odorometer is only partially correct since the instrument to which Allison and Katz applied the term measures only those odors which it itself produces. There is yet no instrument capable of measuring odors, that is, concentrations of molecules of a so-called odorous substance in the air, or of measuring the response of an organism to odors. As stated above, what is actually needed is a means of observing the character of an organism's response to odors of known concentration.

The earliest practical instrument devised was Zwaardemaker's olfactometer which purported to measure olfactory acuity in human beings. Essentially, it consisted of one glass tube within another the walls of which were coated with the substance being tested. The farther the internal tube was withdrawn, the greater the surface of the coated tube exposed to air drawn through by inspiration when held to the nostril, and hence the stronger the odor since intensity was considered to be proportional to the surface exposed for evaporation. The inner tube was graduated to read in purely arbitrary units called olfacties (Fig. 20). However crude and inaccurate this device, it represented the first real step in a difficult direction.

Y-tube Type Olfactometers. Such an instrument was obviously unfit for use with any creature but man. In 1907 a new principle was introduced by Barrows for the study of insect olfaction. This is the principle of the Y-tube. Its introduction ushered in an era of

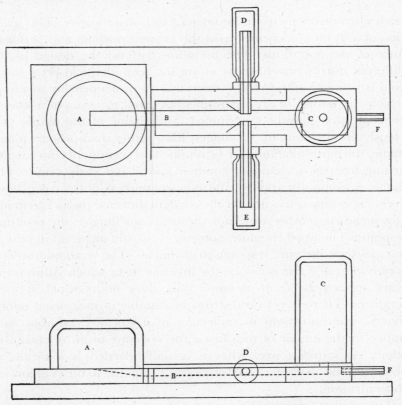

FIG. 21. Top and side views of Barrows' olfactometer. Flies in chamber (A) travel in response to light along trough (B) in the direction of collection chamber (C). In so doing they pass the entrances of two traps (D) and (E), one of which contains an odorous substance. An air current is maintained by suction at F. (Redrawn after Barrows, *J. Exp. Zool.*, 1907.)

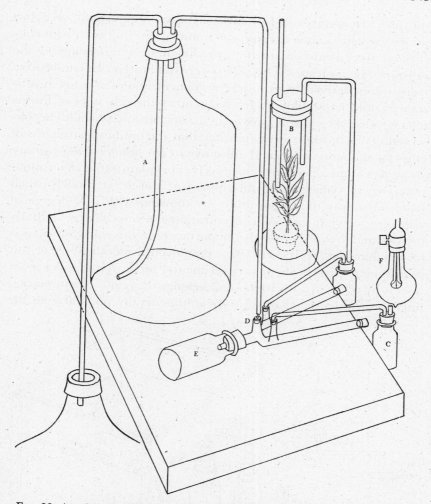

FIG. 22. An olfactometer for testing the reactions of potato beetles and tussock moth larvae. Siphon (A) sucks air from a chamber (B) containing a living plant and from the atmosphere (C) through Y-tube (D). Insects in bottle (E) reacting to light (F) enter the arms of the Y-tube whence they are removed and counted. (Redrawn after McIndoo, *J. Econ. Entomol.*, 1926.)

comparative or qualitative olfactometers some of which, however, could be used semiquantitatively. The principle of the Y-tube is very simple. In Barrows' instrument (Fig. 21) odorless air was allowed to diffuse through one arm in the direction of the stem; odorous air from a trap containing the substance being tested, through the other. Insects passing into the stem in response to light were thus presented with a choice. From the number entering each trap conclusions were drawn as to the attractiveness or repellency of the substance being tested. This was purely qualitative. In consecutive trials odors derived from different concentrations of a given substance could be employed. In this manner the responses of insects to a graded series of approximated concentrations could be observed. Thus did Barrows determine that the optimum strengths of ethyl alcohol and acetic acid attractive to *Drosophila melanogaster* are 20 per cent and 5 per cent respectively. For quantitative determinations the chief objection to this apparatus and to the models which followed it was the impossibility of knowing with any degree of certainty to what gaseous concentration the insects were actually responding. Nevertheless, the importance of this early work was far-reaching for McIndoo (1926) employed a modification of this technique to present the first experimental proof that plants attract insects by odors. To McIndoo also belongs the credit of stimulating interest in instruments of this type and generally popularizing the

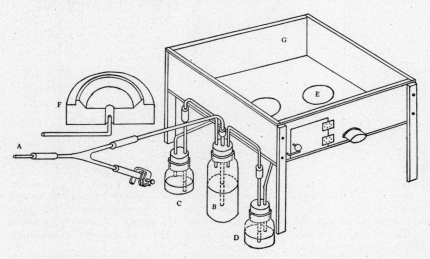

Fig. 23. McIndoo's olfactometer for testing the reactions of blowflies. Air blown through tube (A) is bubbled through water bottle (B). It then passes through control bottle (D) and a bottle (C) containing an odorous substance. The two streams of air emerge through the perforated iron disks (E). This exit mechanism is shown in section at F. Flies liberated in the box (G) (wire screen in cover not shown) respond to appropriate odors emerging at the disks. (Redrawn after McIndoo; *J. Agr. Research*, 1933.)

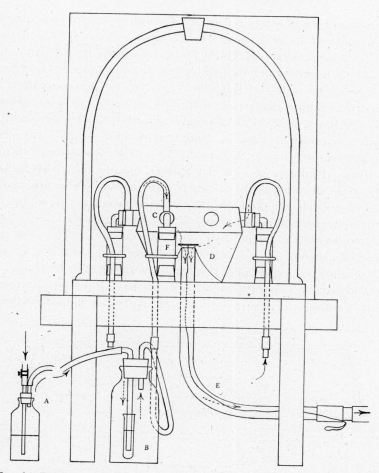

Fig. 24. Ripley's and Hepburn's olfactometer for testing the reactions of fruit flies. Air enters water bottle (A), passes to saturator (B) where it becomes odor-laden, passes via valve bottle (C) to central chamber (D), and leaves by tube (E). Flies are introduced through tube (E). From the central chamber they enter one of eight valve bottles (each connected to a saturator) in response to odor and passing through the valve are retained in trap (F). (Redrawn after Ripley and Hepburn, *Union S. Africa Dep. Agr., Entomol. Mem.,* 1929.)

subject to entomologists. The most important improvement over Barrows' instrument was the addition of a siphon arrangement to suck air through the Y-tube not only from solutions but from living plants as well, so that odors could be disseminated by circulation rather than by diffusion (Fig. 22). In addition to possible errors arising from lack of temperature and humidity control this innovation introduced air velocity as a new source of error. By taking due cognizance of possible sources of error, however, the experimenter was able to derive significant results from his instrument. This olfactometer served well the purpose for which it was designed. It showed that potato beetles are attracted to water extracts, aerated steam distillates, and odors of potato plants, that they are slightly attentive to other Solanaceae, and that tussock moth larvae (*Hemerocampa leucostigma* S. & A.) are slightly attracted to water extracts and steam distillates of elm. In a later modification (McIndoo, 1933) some drawbacks of the early instrument were eliminated. The new model, adapted for experimentation with blowflies, allowed for some control of temperature and humidity, permitted more freedom of movement on the part of the insects, and also allowed the insects more than a single choice (Fig. 23). The advisability of allowing experimental insects several choices between test odor and control is questionable. A single irrevocable choice is obviously a source of error in that average behavior is not obtained.

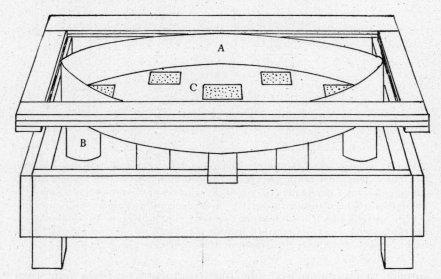

Fig. 25. An olfactometer for testing the olfactory reactions of wireworms. Insects are freed in central chamber (A) (screen in cover not shown). Odorous substances are volatilized by drawing air through tubes (B) which are open at both ends. During tests diffusion carries odor through perforated covers (C) at which place wireworms show response. (Redrawn after Lehman, *J. Econ. Entomol.*, 1932.)

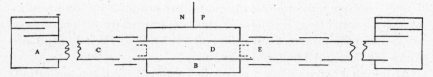

FIG. 26. Diagram of Folsom's chemotropometer: (A) air box, (B) dark box, (C) air tube, (D) reaction tube, (E) reagent tube, (N) side at which insects gather if response is negative, (P) side to which insects go if response is positive. (Redrawn after Folsom, *J. Econ. Entomol.*, 1931.)

On the other hand, it has been proved that conditioning to odors occurs and may affect responses. This apparatus was, nevertheless, used successfully to indicate that fermented casein and baker's yeast, putrid meat, and putrid eggs are exceptionally attractive to blowflies.

Ripley and Hepburn (1929) designed an instrument consisting of a circular chamber with eight holes equally spaced around the wall, each leading to a tube through which air carrying an odor was drawn. Circulation was maintained by withdrawing air through a hole in the bottom of the chamber (Fig. 24). Traps prevented insects from returning to the central chamber, and illumination was central and overhead. Essentially, this was an apparatus for "concentrated trapping" under controlled conditions, a multiplication of the Y-tube principle, and subject to the same limitations. Lehman (1932) also used a multiple tube olfactometer (Fig. 25) for extensive experimentation with wireworms. Here, however, substances were first volatilized by drawing air through the tubes which were open at the ends. During tests diffusion was relied upon to bear odors to the wireworms. Both of these instruments were of the comparative type. In such a capacity they indicated the relative attractiveness of hundreds of chemicals tested. Another successful modification of the Y-tube type olfactometer was that designed by Folsom (1931) (Fig. 26) and termed a chemotropometer. It eliminated the use of light stimuli and active air circulation. It was a nearly completely closed system relying entirely upon diffusion. With it Folsom was able to compare the responses of Mexican boll weevils (*Anthonomus grandis* Boh.) to different parts of the cotton plant, different varieties of cotton, cotton dew, different concentrations of trimethylamine, and different concentrations of ammonium hydroxide. Some of his results are tabulated in Tables 6, 7, and 8.

Table 7 is especially interesting as an illustration of the relation between attractiveness, repellency, and concentration. Other successful Y-tube olfactometers were those of Marshall (1935), Thorpe and Jones (1937), and Hepburn (1943).

Although U-shaped the apparatus designed by Eagleson (1939) (Fig. 27) for use with muscoid flies was fundamentally a Y-tube olfactometer on a large scale. As in most instruments of this type, a light stimulus was employed to cause flies to congregate at the point where they might choose between the test odor and the control.

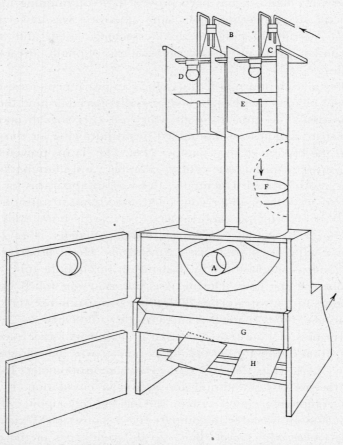

FIG. 27. Eagleson's olfactometer for use with muscoid flies: (A) entrance for introduction of insects, (B) entrance of air stream, (C) flask containing test material, (D) light, (E) light diffusion septum, (F) movable screen partition, (G) window, (H) mirror. (Redrawn after Eagleson, *Soap*, 1939.)

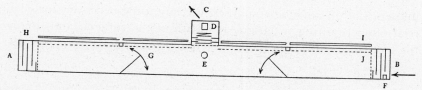

Fig. 28. An olfactometer for testing the reactions of corn earworm moths: (A and B) air inlets, (C) air outlet, (D) fan, (E) moth entrance, (F) attractant, (G) door, (H) light baffles, (I) light proof covers, (J) screen covers. (Redrawn after Ditman, Secrest, and Cory, *Maryland Agr. Exp. Sta. Bull.*, 1941.)

Ditman, Secrest and Cory (1941) designed an apparatus to be used with corn ear worms. It is shown in Fig. 28. More recently Ingle (1943) designed a piece of apparatus which he designated a chemotropometer (Fig. 29). It did not differ in principle from others except that a blue rather than a white light provided the initial stimulus.

The Y-tube olfactometers, admirable within their limitations, suffered many more or less serious defects. They permitted merely of qualitative results or, by the use of repeated runs, semiquantitative results. Results obtained with them were frequently in disagreement with field tests. Snapp and Swingle (1929) found that certain odorous chemicals characteristic of peach, as well as distillates of fruit and bark, were attractive to peach insects in the field but actually acted as repellents when tested in an olfactometer. This difficulty, of course, resolved itself around concentration. The time had come for an instrument in which accurate quantitative tests could be run by means of which one could ascertain threshold values. To be reliable, such an instrument would also have to approximate as nearly as possible the requirements of an ideal olfactometer as enunciated by Hoskins and Craig (1934).

1. The environment should be as nearly normal as possible.
2. A large fraction of the individuals tested should respond.
3. Results should be clearly attributable to chemical stimuli only.
4. Results should be quickly obtainable.
5. Results should be obtainable with either a homogeneous or a heterogeneous population.

Some objections to earlier models were the impossibility of controlling temperature, humidity, etc., in the absence of a normal environment, the small fraction of total individuals responding, the influence of phototactic responses, the long time necessary for each trial often resulting in fatigue or death of the subject, and finally the fact that with many models each animal was given but one op-

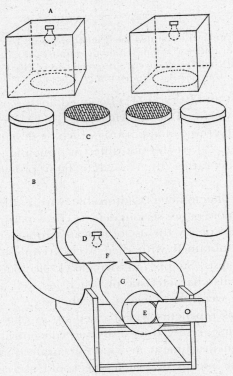

FIG. 29. Ingle's chemotropometer. Flies are drawn into chamber (E) by a light at (D). Arm (F) is separated from E by a screen. Light (D) is then turned off and lights in terminal chambers (A) are turned on, attracting flies from (G) up through arms (B) to screens (C). A fan at F sucks air down both arms. (Redrawn after Ingle, *J. Econ. Entomol.*, 1943.)

portunity to exercise a choice. To meet these objections Hoskins and Craig designed a new type of insect olfactometer.

The prototype of Hoskins' and Craig's olfactometer was designed much earlier by Allison and Katz (1919) for use in investigations of stenches and odors for industrial purposes. Though these new type instruments were actually elaborations of Y-tube models, they owe their success to incorporation of the Venturi tube and may be referred to as Venturi type olfactometers.

Venturi Type Olfactometers. Essentially this kind of instrument delivers an odor of known concentration to a chamber containing the insects. The superiority of this type instrument over earlier ones lies in the fact that the exact concentration of the *odor*, not merely the substance from which it was derived, is known. The Venturi tube is a tube to measure velocity of flow by decrease of pressure. It is nothing more or less than two oppositely directed hollow cones joined together coaxially by a short throat of uniform diameter. While the angular opening of the receiving cone is comparatively large, the mouth of the discharge cone is long and tapering. To measure the velocity of flow, V, let r represent the ratio of the cross-section of the mouth to that of the throat, and the velocity in the latter becomes rV. If h_1, h_2, and h_3 are the heads of the current atmosphere that would give the static, mouth, and throat pressures respectively, and g is gravity acceleration, then

$$V = \sqrt{2g(h_1 - h_2)} = \frac{1}{r}\sqrt{2g(h_1 - h_3)} = \sqrt{\frac{2g(h_2 - h_3)}{r_2 - 1}}$$

Thus to measure V it is merely necessary to connect the mouth and throat to the opposite sides of a manometer. Let p equal the manometer reading and c the value of h per unit of p. Then

$$V = \sqrt{2gcp_1} = \frac{1}{r}\sqrt{2gcp_2} = \sqrt{\frac{2gcp_3}{r^2 - 1}}$$

This instrument is then known as a flow meter. For a detailed exposition of Venturi tubes the reader is referred to Humphreys (1940).

The odorometer of Allison and Katz (Fig. 30) consisted of a number of Venturi type flow meters so arranged that a measured volume of air could be passed at a uniform rate through or over a liquid and this air then mixed with a measured volume of pure air also flowing through at a uniform rate. By determining the loss of weight of liquid and the total volume of air with which this liquid had been mixed it was possible to calculate the concentration in milligrams per liter of air or in parts of vapor per million parts of air.

The Hoskins and Craig apparatus eliminated the accessary flow meters of different capacities as well as the pressure regulators, and

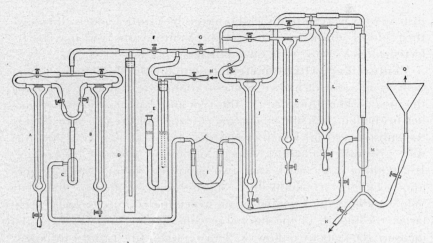

Fig. 30. Allison's and Katz's odorometer: (A) 100-liter-per-hour capacity flow-meter; (B) 3,000 cc. p. hr. cap. flow-meter; (C) emergency water trap; (D) water-pressure regulator for air; (E) mercury-pressure regulator for air; (F) point of regulation to allow passage of small volume of air which goes through meters and saturator; (G) point of regulation to allow passage of large volume of air which subsequently dilutes air from saturator; (H) entrance of air; (I) saturator; (J, K, and L) flow-meters of 50, 500, and 1,200 l. p. hr. cap. respectively; (M) mixer for odorous and pure air; (N and O) exits. (Redrawn after Allison and Katz, *Ind. Eng. Chem.*, 1919.)

added saturating towers filled with glass beads and the testing chambers necessary for use with insects. In short, it consisted of a box on one side of which were two closely adjacent circular holes covered with wire screen. To each, on the outside, came two funnels, one delivering odorless air, the other, a mixture of odorless air and air with which had been mixed a known concentration of the gas being studied. A light attracted the insects, in this case flesh flies, to the ports where they could be counted as they came to rest on either the test or the check area. This apparatus subsequently was modified for use with houseflies (Fig. 31). Aside from the introduction of a heating chamber the new model was substantially similar. Since houseflies did not respond well to light, thermotaxis was relied upon to draw them to the ports. Three variables were anticipated and controlled, temperature, humidity, and air stream velocity. Between 18° and 30° C. there was little attraction of flies to moving air. Between 31° and 41° C. attraction to moving air was appreciable. Heat paralysis set in between 42° and 50° C. Normal activity seemed to prevail at 41° C. Experiments with dry and saturated air versus "ordinary" air indicated a preference for the latter. The most favorable rate of air flow according to criteria of adequate response and normal activity was approximately 7.2 cu. ft. per hour. The concentration of test odors could be determined (1) by determining the loss of weight of liquid when a measured volume of nitrogen was passed through it, this saturated stream then being mixed in desired proportions with pure air, or (2) if complete saturation of the nitrogen is insured, by calculation from the vapor pressure of the liquid. In the latter instance, rigid temperature control is required. This instrument satisfied the requirements for an ideal olfactometer as promulgated by Hoskins and Craig. With it, it was possible to make very precise determinations of the thresholds of response and to ascertain the optimum concentration of ammonia, carbon dioxide, and ethyl alcohol for purposes of attraction (cf. p. 117).

With all of the instruments thus far described, understanding of insects' reactions to test odors has been based upon the only visible manifestation present, the resultant average locomotor behavior of a sample of population when presented with a choice. This indirect approach naturally and unavoidably introduced factors tending to confuse the issue and adversely influence the interpretation of results. Still, it is undoubtedly the most satisfactory way of ascertaining the most attractive of several materials or the most attractive (or repellent) concentration of a given material. And this has been the prime motive behind the construction of Venturi type olfactometers.

There are occasions when it is expedient to know the minimum concentration of a substance which can be detected by a given

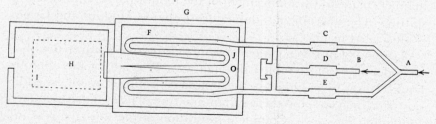

Fig. 31. Diagram of Wieting's and Hoskins' olfactometer designed for use with houseflies: (A) entrance of air, (B) entrance of gas or vapor, (C, D, and E) flowmeters, (F) glass tubing in heating chamber, (G) heating chamber, (H) exhaust port, (I) wire screen cage, (J) light. (Redrawn after Wieting and Hoskins, *J. Econ. Entomol.*, 1939.)

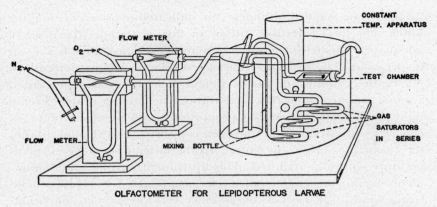

FIG. 32. Olfactometer for use with lepidopterous larvae. (From Dethier, "Laboratory Procedures in Studies of the Chemical Control of Insects," *Am. Assoc. Adv. Sci.*, 1943.)

species, in other words, the threshold. To determine this it is possible to develop other angles of approach to the problem. One way is to discover, where possible, a characteristic response more directly related to olfactory stimulation by which an insect indicates reaction to a test material, i.e., a reaction similar to the proboscis response of flies and butterflies to near threshold concentrations of taste substances. The question of choice is eliminated. Embodying this approach, an olfactometer was designed (Dethier, 1941 and 1943) which rendered possible determinations of the lowest concentrations of certain odorous substances to which lepidopterous larvae gave a visible response. It depended upon the fact that caterpillars respond to odors in a very characteristic and stereotyped manner, namely, by a movement of the mouth-parts reminiscent of spitting. The instrument, adapted from that of Hoskins and Craig, is illustrated in Fig. 32.

A stream of nitrogen gas of constant velocity was saturated with a test odor by being bubbled through a C. P. grade of the odorous liquid. It was then diluted to the desired concentrations by the addition of oxygen gas, and the resultant mixture delivered to a chamber containing the larva. Complete saturation of the nitrogen was insured by the use of three all-glass saturators in series. These were adaptations of the type designed by von Bichowsky and Storch (1915). The flow of gas in each case was regulated by needle valves and measured by glass flow meters (Benton, 1919). The gases were mixed in a chamber which acted also as a valve preventing the more rapidly flowing gas from impeding the flow of the less rapidly flowing one. A larva was then confined in a small glass tube screened with wire at either end to prevent the escape of the animal yet allow free passage through the tube of all gases. This tube was placed within the test chamber to which the mixed gases were delivered and from which they escaped into the room. Saturators, mixing chamber, and test chamber were immersed in a constant temperature bath maintained at 20° C. This temperature was found to be optimum for larvae and convenient for calculations of concentrations. A light suspended over the test chamber illuminated the animal and caused it to remain in a central position where it could be observed with a large magnifying glass. All connections were glass to glass.

Inability to compress or secure compressed air necessitated the use of a cheap, easily obtained gas. Nitrogen filled these requirements. The addition of oxygen is necessary for the maintenance of life. Two variables, the oxygen/nitrogen ratio and the gas velocity, constituted possible sources of error. Carefully controlled experiments indicated that changes in the oxygen concentration of the gas mixture ranging from 20 to 90 per cent oxygen had no measurable effect on the response to odor. The average thresholds for larvae maintained at atmospheric concentrations of oxygen did not differ significantly from those of larvae maintained in a gas mixture containing about 90 per cent oxygen. Therefore, the small changes in oxygen concentration introduced in the course of experiments (80 to 90 per cent) required no further control. Responses to variations in gas velocity accompanied large and abrupt changes only. Thresholds could be determined at different gas mixture velocities since concentrations depended upon the ratio of the rate of flow of oxygen to the rate of flow of nitrogen and not on the total velocity. Thus it was found that the threshold of response to odors remained unaffected by small velocity changes occurring under usual experimental conditions. Concentrations were calculated in terms of grams of solution (for example benzaldehyde) per liter of gas mixture from the following equation:

$$W = \frac{p_a M_a}{\left[(P_b - p_a)RT \right]\left[F_{O_2} - \left(\frac{P_b}{P_b - p_a} \right) F_{N_2} \right]}$$

where W = the number of grams of benzaldehyde per liter of gas mixture, p_a = the vapor pressure of benzaldehyde in millimeters of mercury at the temperature of the solution, M_a = the molecular weight of benzaldehyde, P_b = the barometric reading in millimeters of mercury, R = the gas constant (0.08207 liter atmospheres), T = the absolute temperature, F_{O_2} = the ratio of the rate of flow of oxygen to the rate of flow of nitrogen in liters per minute, and F_{N_2} = the rate of flow of nitrogen in liters per minute. Humidity was not controlled and may have had some slight effect on the threshold values.

As a means of determining threshold values this machine suffered three defects: it was adapted for use with lepidopterous larvae only; it was efficient only with repellents; it gave the threshold of response, not the true threshold.

Methods of Determining Absolute Threshold Concentrations. Two very promising means of determining true threshold values remain. Both are complicated. One relies on conditioned reflexes; the other resorts to the detection of nerve action currents. By conditioning an insect to a given odor so that some recognizable response is elicited every time that odor is present it should be possible to determine threshold values very accurately. The classic experiments involving conditioning were those of von Frisch whereby he trained honey bees to associate odors with food (sugar). Thresholds were not determined, however. Frings (1941) succeeded in conditioning blowflies to the odor of coumarin and associating it with food so that a proboscis response was elicited every time the odor was present. The experiments were directed toward another end, but they point the way to a new approach to the problem of thresholds. Research directed to this end is now in progress. It employs an improved olfactometer incorporating such refinements as rotameters to replace Venturi tubes, fritted glass saturators in lieu of Bichowsky-Storch saturators, magnetic stirrers within the mixing chamber, controlled humidity, and interchangeable mixing chamber-test chamber units to lessen the chance of odor contamination by permitting complete replacement following each test.

The detection of action currents is more direct. By placing electrodes on the antennal nerve it should be possible to pick up electrical impulses when an odor of threshold concentration or greater impinges upon the antennal receptors. The bulk of evidence places the chief olfactory organs of insects in the antennae (see Wiggles-worth, etc.), thus the certainty of working with an appropriate nerve is assured. The obstacles of technique are formidable but not insurmountable. Attempts along these lines were made by Dethier (1941) but were not successful. Nevertheless, a resume of the procedure will indicate the possibilities of the scheme.

Action potentials were recorded photographically by means of a Matthews oscillograph and a resistance-capacity coupled amplifier. A larva to be tested was fastened ventral side down to a block of paraffin by means of fine insect pins. The anterior surface of the head capsule comprising the vertex and adfrontal areas were excised with a pair of iridectomy scissors. Removal of this piece of cuticle leaves a window in the front of the head capsule where the brain and its nerves lie exposed. Due to the shortness of the nerve (1.75 mm.), all recordings were monopolar. A fine silver wire hooked under the nerve raised it into the air. A thicker silver wire led down into the tissues and body juices of the head. This lead was used alternately as ground and grid. Unfortunately, recordings of this nature rendered impossible any interpretation of wave form, and spike heights were purely relative.

Finally, there remains a method by which attractive or repellent materials might be evaluated. This does not yield data on thresholds but is worthy of mention at this time. It consists of exposing organisms to a compound field of excitation involving two opposed stimuli. Many insects orient very stubbornly and in a very precise manner to light. It is possible to cause them to deviate from their phototactically directed path by exposing them simultaneously to another type of stimulus directed from a different quarter. Crozier and Pincus (1927) found that when this added stimulus was gravity there was a mathematical relation between the strength of each stimulus and the angle of deviation from the original path. It is quite certain that odors will cause deviation. If the above procedure could be adapted to olfactory stimulation, it would provide a less arbitrary index of the attractiveness or repellence of substances than those now used. Crombie (1944) has achieved a degree of success in balancing antagonistic stimuli by exposing blowflies (*Calliphora eryth-rocephala* and *Lucilia sericata*) to a compound field of excitation in a Venturi type olfactometer. Rather than attempt to record a deviation of path of freely crawling insects he arranged conditions so that the flies had a choice of crawling along one arm of a Y-tube toward light (they are positively phototactic) against a repellent odor or away from light along a tube devoid of odor. Both stimuli were above threshold. In one series of experiments a uniform intensity of light opposed different concentrations of menthol; in the other,

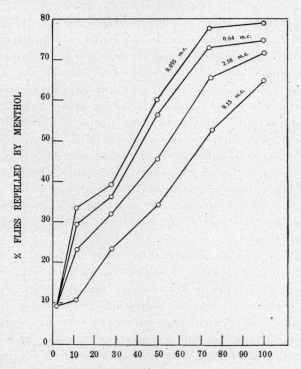

FIG. 33. Variation in response of *Lucilia* adults to antagonistic stimuli of light and odor with light intensity constant and odor concentration varied. (Redrawn after Crombie, *J. Exp. Biol.*, 1944.)

the concentration of menthol was constant and light intensity varied. Figs. 33 and 34 present the results of these experiments. No attempt was made to measure the different intensities of odor and light which would exactly balance each other, leaving 50 per cent of the flies in each arm of the Y. Nevertheless, this appears to be an excellent means of comparing the absolute repellency of different compounds.

The knowledge gained and to be gained in the future by properly operated olfactometers is of extreme importance. Y-tube type instruments still retain a place in modern entomological research. Simply and quickly constructed they serve satisfactorily in preliminary or

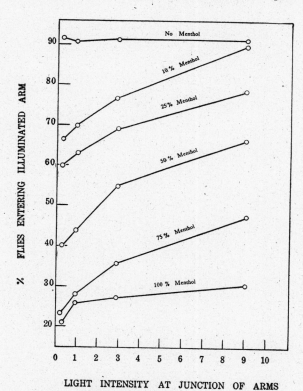

FIG. 34. Variation in response of *Lucilia* adults to antagonistic stimuli of light and odor with odor concentration constant and light intensity varied. (Redrawn after Crombie, *J. Exp. Biol.*, 1944.)

routine experiments intended to screen candidate attractants or repellents. For more exact information of a quantitative nature a Venturi type instrument is indicated. Models currently in use are suited to the determination of rejection thresholds of candidate olfactory repellents. Obviously, a low rejection threshold is a desirable characteristic in any repellent. Perhaps the most valuable information gleaned in the past through the instrumentality of olfactometers

Table 37

THRESHOLDS OF RESPONSE OF CERTAIN INSECTS TO ATTRACTIVE ODOROUS SUBSTANCES

Species and Authority	Skatole	Indole	Ammonia	Ethyl Alcohol	Benzol	Benzaldehyde	Cyclohexane
Necrophorous spp. (Abbott, 1937)	0.00009 g./100 cc. water						
Habrobracon juglandis (Wirth, 1928)				5.0–20.0 mg./l.	0.5–3.0 mg./l.		3.0–5.0 mg./l.
Geotrupes sylvaticus (Warnke, 1931)	0.003 – 0.009 mg./l.	0.003 mg./l.	1 cc. of 0.050%/l. air				
G. vernalis (Warnke, 1931)	0.003 – 0.009 mg./l.	0.003 – 0.009 mg./l.	1 cc. of 0.050%/l. air				
G. stercorius (Warnke, 1931)		0.15 mg./l.					
Musca domestica (Wieting & Hoskins, 1939)			0.04 mg./l. (41° C.)	0.23 g./l. (41° C.)			
Pieris rapae (Dethier, 1941)						580×10^{-7} g./l.	
Malacosoma americana (Dethier, 1943)						435×10^{-7} g./l.	

was that chemotaxis is measurably influenced by the molecular concentration of the stimulating substances involved.

Molecular Concentration. With odor as with any stimulus a certain intensity must be attained before any response is elicited

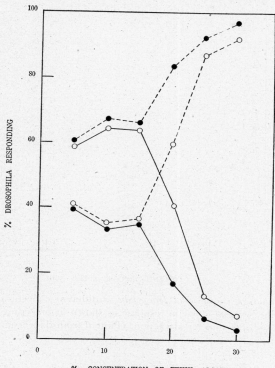

% CONCENTRATION OF ETHYL ALCOHOL

FIG. 35. Olfactory response of *Drosophila* to different concentrations of ethyl alcohol. Solid line indicates the percentage of males (solid circles) and females (open circles) attracted. Broken line indicates the percentage repelled at different concentrations. Note that repellency increases as concentration is increased beyond optimum for attraction. (Plotted from data from Reed, *Physiol. Zool.*, 1938.)

from an organism. In the case of chemical stimuli this intensity is concentration. Odor is an aerial molecular suspension. When the number of molecules of odorous substance in air impinging on a receptor is sufficient to initiate a nerve impulse at that end organ, it is termed the threshold concentration. Usually this is not sufficient

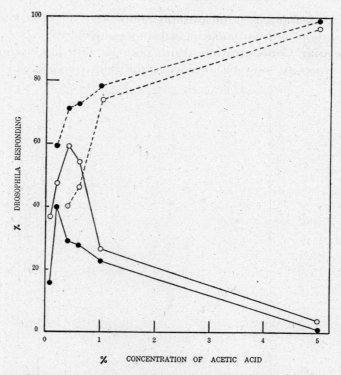

FIG. 36. Olfactory response of *Drosophila* to different concentrations of acetic acid. Note same reversal effect in response as shown in preceding figure as concentration passes optimum for attraction. (Plotted from data from Reed, *Physiol. Zool.*, 1938.)

to evoke any recognizable response from the insect. Most so-called threshold values actually represent the threshold of response. They are presumed to approximate the former values and in the absence of more exact data are informative.

It was long assumed that the olfactory acuity of insects far surpassed that of man. A few figures on threshold of response indicate that such is actually the case (Table 37). As molecular concentration exceeds the actual threshold value a point is reached which may be termed the threshold of response. This reaction may take the form of acceptance or avoidance. Which it is, is predetermined by the genetic constitution of the animal. Herein lies the fundamental distinction between attractants and repellents. The genetically directed response to an odorous substance at this point determines it

as an attractant or a repellent. If a substance is repellent at this concentration, it remains so as the concentration is increased but may
become more strongly repellent. Above a given concentration it
may affect a different group of receptors, those of the common
chemical sense. If it is initially attractive, it becomes more so as
concentration is increased till an optimum is reached. Beyond this
point attractiveness decreases with increased concentration till the
nature of the response is completely reversed. For every attractant
thus far tested there is a concentration in excess of which the substance acts as a repellent. Graphs (Figs. 35 and 36) based on Reed's
(1938) experiments with the olfactory reactions of *Drosophila* illustrate the relation between optimum concentration and reversal
concentration. The graph (Fig. 37) based on Folsom's (1931) experiments with the Mexican boll weevil also illustrates reversal
concentration. Optimum and reversal concentrations are further
illustrated in Figs. 38 and 39 based on the results of Wieting and
Hoskins (1939).

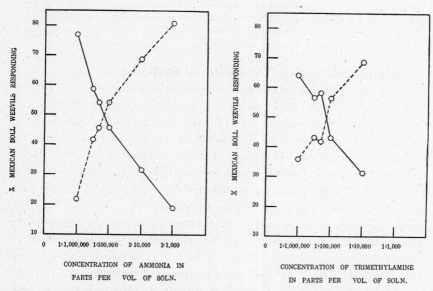

Fig. 37. Olfactory responses of Mexican boll weevils to different concentrations
of ammonia and trimethylamine. Note that as attraction (solid line) decreases
as concentration is increased beyond optimum, repellency (broken line) increases.
Reversal concentration may be said to be that at which 50 per cent of the insects
respond positively and 50 per cent respond negatively. (Plotted from data from
Folsom, *J. Econ. Entomol.*; 1931.)

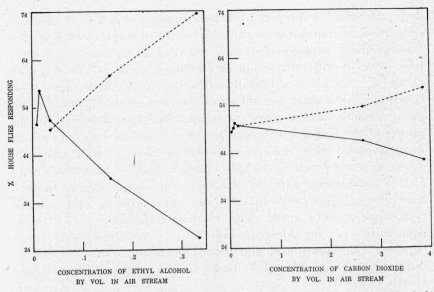

FIG. 38. Olfactory responses of houseflies to different concentrations of ethyl alcohol and carbon dioxide. The reversal concentration is that at which 50 per cent of the flies are attracted (solid line) and 50 per cent are repelled (broken line). (Plotted from data from Wieting and Hoskins, *J. Econ. Entomol.*, 1939.)

Conditions Affecting Thresholds of Response.

In olfactometers three of the most important factors tending to influence chemotactic behavior are temperature, humidity, and air movement. Since most experiments involving the use of such apparatus are undertaken to supply data relative to thresholds or comparative responses to different substances, every effort is made to control variables. Incidental observation in the course of such work has indicated that while chemotactic behavior is affected by temperature, humidity, and air currents, the influence is an indirect one. By affecting the over-all behavior of an insect these factors cause a variation in response that may incorrectly be interpreted as a modification of threshold. Abbott (1932a, 1936, 1938) has shown that temperature and humidity affect the proboscis response of blowflies, and others state that the behavior of various insects in olfactometers is profoundly influenced by the same variables. On the other hand, there is little conclusive evidence that absolute threshold is modified directly by either temperature or humidity changes. Some experiments by Woerdeman (1935), however subjective, indicate that, for

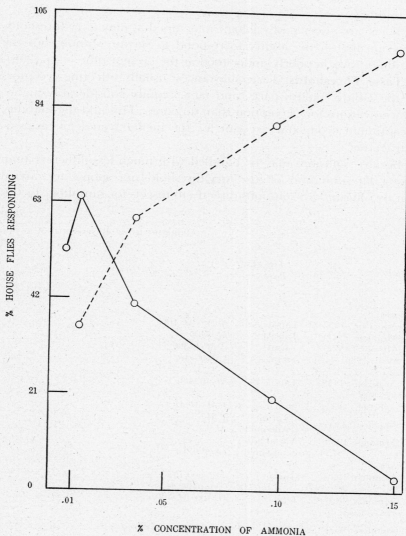

Fig. 39. Olfactory responses of houseflies to different concentrations of ammonia. Note the same reversal effect as shown in preceding figure. (Plotted from data from Wieting and Hoskins, *J. Econ. Entomol.*, 1939.)

human beings at least, odor perception is more acute at low humidities.

There are internal physiological factors that may alter thresholds or even abolish the ability to respond to certain stimuli, but the action of these is poorly understood at the present time (cf. p. 255).

Taste Thresholds. Taste substances, usually affecting receptors in the mouth, mouth-parts, and tarsi, require considerably higher concentrations for perception than do odors. This and the inability to stimulate from afar account for the ineffectiveness of tastes as attractants.

Because solutions may be handled with much less difficulty than gases, determinations of gustatory thresholds of response are carried out by offering test solutions directly to insects for sampling. In this

Table

THRESHOLDS OF RESPONSE OF CERTAIN

Species and Authority	Type of Receptor	Sucrose	Glucose	Fructose
Pyrameis				
(Anderson, 1932)	Tarsal	$M/10$–$M/100$		
(Minnich, 1921)	Tarsal⎫	$M/3200$ to		
(Minnich, 1922)	Tarsal⎭	$M/12800$		
(Weis, 1930)	Tarsal	$1/1600\ M$		
Danais				
(Anderson, 1932)	Tarsal	$M/102400$		
HONEY BEE				
(v. Frisch, 1934)	Oral	$1\ M$–$M/16$		
(Marshall, 1935)	⎧Tarsal	$1\ M$		
	⎩Antennal	$M/12$		
(Minnich, 1932)	Tarsal and antennal	$64/100\ M$		
Calliphora				
(Minnich, 1929)	Tarsal	$1/25600$ to $1/6400\ M$		
(Minnich, 1931)	Oral	$M/100$–$M/400$		
CATERPILLARS				
(Dethier, 1937)	Oral	$0.1\ M$	$0.5\ M$	
(Dethier, 1939)	Oral	0.01–$0.02\ M$	0.02–$0.03\ M$	0.01–$0.06\ M$
(Eger, 1937)	Oral			
(Frings, 1945)	Oral			
ANTS				
(Schmidt, 1938)	Oral	$1/200\ M$ to $1/800\ M$	$1/8\ M, 1/50$ $M, 1/32\ M$	$1/64\ M, 1/50$ $M, 1/32\ M$

manner values of thresholds of response of numerous insects to taste substances have been determined (Table 38). As with odors, no satisfactory methods for determining absolute thresholds have been evolved. Here again studies of action currents and of conditioning probably afford the only practical means of achieving this end. While conditioning to taste substances most certainly involves greater difficulties than conditioning to odors, it has in some instances been successfully accomplished by reward and punishment. *Dytiscus*, a water beetle, has been thus conditioned by being fed after stimulation with salt and denied food after stimulation with sugar. Conceivably, insects might also be conditioned to certain tastes by electric shock, but experiments of this sort have not been reported.

The phenomenon of response reversal at critical concentrations

38

INSECTS TO TASTE SUBSTANCES

Lactose	Maltose	NaCl	HCl	Quinine Salts
			1 M	$M/600$ $1/32000\ M$
		$N/16$		$N/80$
64/100	64/100			
		$M/8$ 0.89 N	$\begin{cases}0.0005\ \text{to}\\0.005\ M\end{cases}$ $\begin{cases}0.003–0.006\ M\\0.004–0.005\ M\end{cases}$ $M/320$ 0.083 N	$M/10240$
0	1/100 M			

above absolute threshold obtains with taste substances too. Some acids acceptable at low concentrations become repellent at higher concentrations. The impossibility of maintaining acceptable sugars in solution at sufficiently high concentration may account for their failure to incite negative responses.

An effect of temperature and humidity on taste thresholds has not been demonstrated, but diet, age, and other factors are known to complicate the accurate determination of thresholds.

REFERENCES

Abbott, C. E.: The proboscis response of insects, with special reference to blow-flies, *Ann. Entomol. Soc. Am.*, **25**(1), 241–244 (1932).

——: The effect of temperature and relative humidity upon the olfactory responses of blowflies, *Psyche*, **39**(4), 145–149 (1932a).

——: Local irritation and the olfactory response, *Bull. Brooklyn Entomol. Soc.*, **31**(5), 203 (1936).

——: The physiology of insect senses, *Entomologica Am.*, **16**(4), 225–280 (1937).

——: The effects of temperature and humidity upon the proboscis response of the blowfly, *Lucilia sericata* Meig., to odorous substances, Abstr. Doctors' Dissertations, No. 25, Ohio State University Press, 18 pp., 1938.

Allison, V. C., and S. H. Katz: An investigation of stenches and odors for industrial purposes, *Ind. Eng. Chem.*, **11**(4), 336–338 (1919).

Anderson, A. L.: The sensitivity of the legs of common butterflies to sugars, *J. Exp. Zool.*, **63**(1), 235–259 (1932).

Barrows, W. M.: The reactions of the pomace fly, *Drosophila ampelophila* Loew, to odorous substances, *Ibid.*, **4**(4), 515–537 (1907).

Benton, A. F.: Gas flow meters for small rates of flow, *Ind. Eng. Chem.*, **11**(7), 623–629 (1919).

von Bichowsky, F. R., and H. Storch: An improved form of gas-washing bottle, *J. Am. Chem. Soc.*, **37**(12), 2695–2696 (1915).

Chaplet, A.: Les olfactomètres et l'olfactométrie, *Parfumerie moderne*, **30,** 3–13 (1936).

Crombie, A. C.: On the measurement and modification of the olfactory responses of blowflies, *J. Exp. Biol.*, **20**(2), 159–166 (1944).

Crozier, W. J., and G. Pincus: On the equilibration of geotropic and phototropic excitations in the rat, *J. Gen. Physiol.*, **10**(3), 419–424 (1927).

Dethier, V. G.: Gustation and olfaction in lepidopterous larvae, *Biol. Bull.*, **72**(1), 7–23 (1937).

——: Taste thresholds in lepidopterous larvae, *Ibid.*, **76**(3), 325–329 (1939).

——: The function of the antennal receptors in lepidopterous larvae, *Ibid.*, **80**(3), 402–414 (1941).

——: Testing attractants and repellents, "Laboratory Procedures in Studies of the Chemical Control of Insects," Washington, D.C., Am. Assoc. Adv. Sci., 1943, pp. 167–172.

Ditman, L. P., J. P. Secrest, and E. N. Cory: Studies on corn ear worm control, *Maryland Agr. Exp. Sta., Bull.* 439, 205–223 (1941).

Eagleson, C.: Insect olfactory responses, *Soap*, **15**(12), 123–127 (1939).

Eger, H.: Über den Geschmackssinn von Schmetterlingsraupen, *Biol. Zentr.*, **57**(5/6), 293–308 (1937).

Folsom, J. W.: A chemotropometer, *J. Econ. Entomol.* **24**(4), 827–833 (1931).

Frings, H.: The loci of olfactory end-organs in the blowfly, *Cynomyia cadaverina* Desvoidy, *J. Exp. Zool.*, **88**(1), 65–93 (1941).

——: Gustatory rejection thresholds for the larvae of the cecropia moth, *Samia cecropia* (Linn.), *Biol. Bull.*, **88**(1), 37–43 (1945).

von Frisch, K.: Ueber den Geschmackssinn der Biene. Ein Beitrag zur vergleichenden Physiologie des Geschmacks, *Z. vergl. Physiol.*, **21**(2), 1–156 (1934).

Hepburn, G. A.: A simple insect cage-olfactometer, *Onderstepoort J. Vet. Sci. Animal Ind.*, **18**(1), 7–12 (1943).

Hilton, W. A.: Sense organs. XIII. Olfactory receptors, *J. Entomol. Zool.*, **35**(3), 39–48 (1943).

Hoskins, W. M.: Recent contributions of insect physiology to insect toxicology and control, *Hilgardia*, **13**(6), 307–386 (1940).

——, and R. Craig: The olfactory responses of flies in a new type of insect olfactometer, *J. Econ. Entomol.*, **27**(5), 1029–1036 (1934).

Humphreys, W. J.: "Physics of the Air," 3d ed., New York, McGraw-Hill Book Co., 1940.

Ingle, L.: An apparatus for testing chemotropic responses of flying insects, *J. Econ. Entomol.*, **36**(1), 108–110 (1943).

Lehman, R. S.: Experiments to determine the attractiveness of various aromatic compounds to adults of the wire worms, *Limonius* (*Pheletes*) *canus* Lec. and *Limonius* (*Pheletes*) *californicus* Mann., *Ibid.*, **25**(5), 949–958 (1932).

McIndoo, N. E.: An insect olfactometer, *Ibid.*, **19**(3), 545–571 (1926).

——: Olfactory responses of blowflies, with and without antennae, in a wooden olfactometer, *J. Agr. Research*, **46**(7), 607–625 (1933).

Marshall, J.: On the sensitivity of the chemoreceptors on the antenna and foretarsus of the honey bee, *Apis mellifica* L., *J. Exp. Biol.*, **12**(1), 17–26 (1935).

Minnich, D. E.: An experimental study of the tarsal chemoreceptors of two nymphalid butterflies, *J. Exp. Zool.*, **33**(1), 173–203 (1921).

——: The chemical sensitivity of the tarsi of the red admiral butterfly, *Pyrameis atalanta* Linn., *Ibid.*, **35**(1), 57–81 (1922).

——: A quantitative study of tarsal sensitivity to solutions of saccharose, in the red admiral butterfly, *Pyrameis atalanta* Linn., *Ibid.*, **36**(4), 445–457 (1922a).

——: The chemical sensitivity of the legs of the blow-fly, *Calliphora vomitoria* Linn., to various sugars, *Z. vergl. Physiol.*, **11**(1), 1–55 (1929).

——: The sensitivity of the oral lobes of the proboscis of the blowfly, *Calliphora vomitoria* Linn., to various sugars, *J. Exp. Zool.*, **60**(1), 121–139 (1931).

——: The contact chemoreceptors of the honey bee, *Apis mellifera* Linn., *Ibid.*, **61**(3), 375–393 (1932).

Reed, M. R.: The olfactory reactions of *Drosophila melanogaster* Meigen to the products of fermenting banana, *Physiol. Zool.*, **11**(3), 317–325 (1938).

Ripley, L. B., and G. A. Hepburn: A new olfactometer successfully used with fruit flies, *Union S. Africa, Dep. Agr., Entomol. Mem.* 6, 55–74 (1929).

Schmidt, A.: Geschmacksphysiologische Untersuchungen an Ameisen, *Z. vergl. Physiol.*, **25**(3), 351–378 (1938).

Snapp, O. I., and H. S. Swingle: Further results with the McIndoo Insect Olfactometer, *J. Econ. Entomol.*, **22**(6), 984 (1929).

Thorpe, W. H., and F. G. W. Jones: Olfactory conditioning in a parasitic insect and its relation to the problem of host selection, *Proc. Roy. Soc. London*, **124** (B834), 56–81 (1937).

Warnke, G.: Experimentelle Untersuchungen über den Geruchssinn von *Geotrupes sylvaticus* Panz. und *Geotrupes vernalis* Lin. *Z. vergl. Physiol.*, **14**(1), 121–199 (1931).

Weis, I.: Versuche über die Geschmacksrezeption durch die Tarsen des Admirals, *Pyrameis atalanta* L., *Ibid.*, **12**(2), 206–248 (1930).

Wieting, J. O. G., and W. M. Hoskins: The olfactory responses of flies in a new type of insect olfactometer, *J. Econ. Entomol.*, **32**(1), 24–29 (1939).

Wirth, W.: Untersuchungen über Reizschwellenwerte von Geruchsstoffen bei Insekten, *Biol. Zentr.*, **48**(9), 567–576 (1928).

Woerdeman, H.: L'influence de la teneur en vapeur d'eau d'un gaz odorant sur la sensation olfactive, *Arch. néerland. physiol.*, **20**(4), 591–595 (1935).

Zwaardemaker, H.: "Die Physiologie des Geruchs," Leipzig, Wm. Engelmann, 1895.

Chapter 7

Baits and Traps

Application of Basic Principles. Notwithstanding the increasing amount of interest in attractants, application of the principles in this field to economic problems still enjoys but a minor rôle in economic entomology. Baits to be effective must be attractive. Attractants for baits are still evolved from long arduous tests with hundreds of miscellaneous chemical compounds. Application of certain fundamental principles and understanding of the intricacies of chemoreception would eliminate much undirected and repetitious work, and by directing efforts into narrower channels, would tend to produce greater results. Studies have proved the preëminence of feeding-type attractants over others and the necessity of not confusing different types. They have outlined the broad categories of odorous materials to which different groups of insects are normally exposed. They have shown the importance of investigating some related compounds and not others, the preference of certain species for certain molecular combinations, the great importance of concentration and mixtures, and the value and shortcomings of each method of assaying prospective attractants. The relation of some of these more fundamental aspects to specific problems is presented by way of illustration in this chapter.

Uses of Attractants. Viewing the uses to which attractants may be put one realizes the embryonic nature of the subject at this point and the need for advancement. The more important uses are obvious: (1) to lure insects into traps or to poisons to decimate the population; (2) to sample local populations; (3) to offset the repellent properties of certain sprays; (4) to lure insects away from crops; (5) to act as counteragents against which to test the efficiency of repellents.

1. The hope behind the early development of attractants was that local populations could be so drained that a species could no longer be considered as a pest. The attractant is used to lure insects to traps or to poison. These initial hopes have not as yet been entirely fulfilled. Scott and Milam (1942 and 1943) have developed a technique for the control of tobacco and tomato hornworms to the

point where infestations can be materially reduced. This is one of the more promising cases. In the case of the Japanese beetle, traps are of some value in reducing population density in areas in which the species has become established (Fleming, Burgess, and Maines, 1940). Undoubtedly they are most valuable in localities distant from a generally infested region, especially during the ten-year period immediately following the beetle's introduction. Such a period generally elapses before beetles become sufficiently well established to constitute a major infestation. At this time the develop-

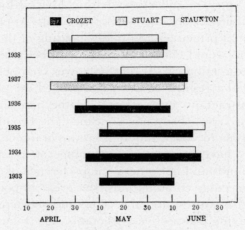

FIG. 40. Emergence of spring brood adults of the codling moth at three locations in Virginia. (Redrawn after Bobb, Woodside, and Jefferson, *Virginia Agr. Exp. Sta. Bull.*, 1939.)

ment of infestation may be retarded by the extensive employment of traps. The value of traps in reducing the population in *old* infested areas is debatable. It is still a subject of experimentation under critical survey and awaiting further development. At the present time trapping relegated to this end is an adjunct to other means of control.

2. Whereas baited traps as sole means of control are not at the moment eminently successful in reducing widespread infestations, they serve well as a means of sampling insect populations. They are of value in determining the presence of suspected invaders in a new area. Federal and state agencies use thousands of traps each year to

capture pioneer Japanese beetle invaders in remote localities and thus prevent the establishment of serious infestation. In this connection traps will attract beetles even when diligent search has previously failed to reveal their presence.

With traps one can also calculate population density as an index of the efficiency of whatever control measures are in force at the time. Peak breeding seasons can be determined for different species at the time of emergence of different broods, thus ascertaining the proper time for control spraying. For example, Worthley (1932) found by the use of bait pails that spraying against codling moth should follow within 10 days of the first peak catch of moths (the approximate incubation period of the egg) and 10 days after each other peak except when peaks are less than 10 days apart. The records obtained by the use of traps in this respect present an accurate picture (Fig. 40) of codling moth activity in an orchard (Bobb, Woodside, and Jefferson, 1939; Eyer and Medler, 1940). In the case of mosquitoes, traps are used to analyze local populations as a prelude to the adoption of mosquito control measures and then to supply data regarding the efficiency of the abatement program (Dethier and Whitley, 1944).

3. For their effect, stomach insecticides depend upon the quantity ingested. The addition of an attractant to such sprays serves to encourage ingestion and offset repellent qualities that the spray might possess. This phase of work is directed against feeding forms which in many insects are the larval stages. No attractant has been found in any of the cases studied which will cause increased feeding though there are preferred taste substances which show promise. Siegler (1940) designated as "larval attractants" those substances which when combined with stomach poisons enticed larvae to the poison for feeding. After testing carbohydrates, polyhydroxy alcohols, and dibasic hydroxy acids he found that on the basis of increased feeding by hatching codling moth larvae brown sugar increased the toxicity of lead arsenate, calcium arsenate, nicotine, bentonite, and phenolthiazine. Sorbitol, sucrose, corn sirup, d-fructose, glycerol, and malic acid did the same with lead arsenate. Molasses increased the toxicity of nicotine, bentonite, and phenolthiazine. Brown sugar was ineffective with derris and pyrethrum. Siegler and Jones (1942) then experimented with extracts of apples as adjuvants to lead arsenate against codling moth larvae. Testing

alcoholic extracts by the apple plug method and using the nonvolatile portion as a basis of comparison, they found that extracts
with a high percentage of nonvolatile material increased the attractiveness of poisoned food. It was concluded that this was due to the
sugars present. Brown (1942) and Wingo and Brown (1942) confirmed that brown sugar acted in a similar capacity. Ditman,
Secrest, and Cory (1941) found that mixed poison and cane sugar
sprayed on the entire corn plant was effective against corn ear worm
moths. No suitable olfactory attractant was found. The pepper
maggot fly also has been reported (Burdette, 1932) as reacting
positively to sprays containing invert sugar. Munger (1942) reported successful control of citrus thrips with tartar emetic-sugar
sprays.

. From the foregoing it is clear that effective substances in every
case are sugars. These are taste substances (gustatory stimuli) and
as pointed out in Chapter 2, unlike olfactory stimuli, are not true
attractants. True larval attractants usually are essential oils (cf.
Chapter 3).

4. Theoretically it is quite within the realm of possibility to develop artificial attractants to compete against nature to such an
extent as to lure pests away from crops, not so much to decimate the
population as to permit the crops to mature with impunity in the
absence of their parasites. The nearest approach to this achievement
is the frequently employed experimental practice of interspersing
rows of a given crop with rows of a preferred plant. According to
some reports such procedures cause a reduction in the number of
insects infesting the crop. For example, the planting of strips of
maize on the windward side of cotton reportedly reduces the jassid
count on the cotton (Gaddum, 1942).

5. The most commonly employed method of evaluating repellents is that of determining the ability of the test material to counteract
a powerful natural or artificial attractant. With respect to blow-
fly-repellents this technique has been employed by many investigators (cf. p. 220). The experiments of Loeffler and Hoskins (1943)
will illustrate the principle involved. These authors found that
gravid flies were drawn to sheep for oviposition because of specific
attractants operative only under abnormal conditions. Blood proved
to be the most powerful of these. It was superior to ground larvae,
larval excrement, and Hobson's indole-ammonium carbonate mix-

ture. Maximum attractiveness was attained after a 48-hour period of exposure at 38° C. Not only was sheep, cow, horse, and hog blood equally effective, but the stimulus operated independently of any specific factor associated with sheep. By attaching a piece of blood-soaked cotton beneath the wool it was possible to make sheep uniformly attractive to fly blow. Then by spraying half of a flock of treated sheep with repellent and retaining the remaining half as controls the effectiveness of the repellent could be determined by its ability to overcome the attractant and so prevent fly blow.

The trend in the evolution of attractants is from natural to synthetic substances. Where traps are baited with the former they are no more powerful than the same attractant normally present in the habitat. Since a bait must compete with nature, it should, to be effective, possess a better attractant. Thus chemical compounds are substituted for natural attractants to provide a more powerful stimulus though at the present time no synthetic attractant has been developed that is superior to favored foods. This is even true for highly successful Japanese beetle baits (Langford, Muma, and Cory, 1943). The most highly evolved bait is one composed of pure chemical compounds mixed according to standardized formulae.

FIG. 41. Metal trap baited with sex attractant and used to capture male gypsy moths. (Courtesy, U.S. Department of Agriculture, Bureau of Entomology and Plant Quarantine.)

Of the types of attractants outlined in Chapter 1 that could be used in baits, feeding-type attractants are superior to all. This is to be expected in view of the fact that feeding-type attractants are all-inclusive while sex or ovipositional-type attractants affect at most 50 per cent of the population, the males in the former case, the females in the latter.

Poison Baits. When attractants are used in traps or in conjunction with poison, they are termed baits. Actually the attractant may be the bait or may be merely one component of the bait. When used in traps the attractant is frequently used alone, hence is the bait. This is the arrangement commonly set up to catch flying insects. Nonflying insects such as grasshoppers, crickets, and larval forms, especially cutworms, are killed with poison baits of which the attractant is merely one component. Such baits are all basically similar, consisting of a carrier (bran, sawdust, chaff, dry manure, etc.) to give body, an attractant (an odorous compound such as amyl acetate added to a palatable ingredient such as molasses), and a poison (usually an arsenical). Sometimes the use of an attractant does not enhance a bait sufficiently to justify its added cost (Walton and Whitehead, 1945).

Trap Baits. For use with flying insects traps generally replace poison. The principle behind the use of the attractant remains the same. In the entire procedure the most important and difficult phase is the selection and development of a proper attractant. Success in this direction depends upon a thorough appreciation and understanding of the principles enunciated in the foregoing chapters. Some measure of success has been attained with a select number of economically important species chief of which are the Japanese beetle, codling moth, oriental fruit moth, tobacco hornworm, fruit flies, and mosquitoes.

JAPANESE BEETLE. To date the best attractants discovered for the Japanese beetle are components of essential oils (cf. p. 71), namely, geraniol obtained commercially by the distillation of oil of citronella, and eugenol, one of the constituents of clove oil. Since none of the commercial geraniols are chemically pure, the various grades differ in degree of attractiveness. Optimum attractiveness is obtained when the following specifications are met:

Specific gravity at 20° C. 0.875–0.895
Total free alcohols as geraniol
   and citronellol . . . . Not less than 70%
Ester content . . . . Not less than 15%
Aldehydes as citronellal . . . . Not more than 3.5%
Solubility . . . . 1 part in 2 parts 70%
   alcohol
Boiling range . . . . Not more than 5% to
   distill below 255° C. or
   more than 18% above
   245° C.

Odor. . . . . Absence of any significant indication that for-
eign materials have
been added

Eugenol derived from clove oil and meeting the following specifications is most satisfactory:

Specific gravity. 1.064–1.070
Solubility. . . . . 1 part in 2 parts 70%
alcohol
Boiling range. . . . . 250–255° C.
Odor and color. . . . . colorless to pale yellow
thin liquid with a strong
aromatic odor of cloves
and a pungent spicy
taste

Recently Fleming and Chisholm (1944) have shown that anethole
may be substituted for geraniol and pimenta leaf oil for eugenol.
Langford and Cory (1946) found that caproic acid and phenyl
ethyl butyrate are effective attractants.

MIXTURES. More attractive than either compound alone is a mixture of geraniol and eugenol 10 to 1 parts by volume respectively.
This follows one of the most important principles relating to attractants—that mixtures are superior to any one of their component
parts (Richmond, 1927; Van Leeuwen and Metzger, 1930; Ripley
and Hepburn, 1931; Hutner, Kaplan and Enzmann, 1937; Fleming, *et al.*, 1940). The reason for this lies in the fact that attractants
are odors, and what has previously been said of plant odors in this
regard holds true for all natural attractants: The odor is the result
of a mixture and blending of incredibly minute quantities of many
chemicals even though usually one chemical may be fundamentally
responsible for the odor (Dethier, 1941). Eyer and Medler's (1940)
study of codling moth attractants indicated that no one chemical
was the most important or key factor in the chemotactic response of
this insect. Working with the same insect Van Leeuwen (1935 and
1943) found that combinations of attractants were more effective
than any one constituent alone. Pine tar oil was only one-half as
attractive alone as when combined with standard bait. Hutner,
Kaplan and Enzmann stated that for *Drosophila* artificially blended
mixtures of attractive chemicals are not badly inferior to natural

baits. No one chemical has yet been found that is strongly attractive to both sexes of *Pterandrus*, and Ripley and Hepburn believe that no one attractive chemical exists because (1) it would either have to be common to all fruits or numerous in kind, and (2) it would have to occur in appreciable quantities so as not to be obscured. Had this been true, they argued, the compound would have been discovered in the course of the hundreds of tests conducted to locate it. All evidence points to the agency of a mixture. They are of the opinion that powerful artificial attractive mixtures can be duplicated.

CODLING MOTH. Different workers have developed various successful baits for use with the codling moth. All are feeding-type attractants and are attractive basically because of fermentation products elaborated by bacteria through contamination from the air (cf. p. 81).

Stock sirup . . . . 1 part
Water . . . . 20 parts
Anethole . . . . 1 cc. per qt. soln.
                (Bobb, Woodside, and Jefferson, 1939)

Brer Rabbit Green Label
    molasses . . . . 1 part
Water . . . . 9 parts
Anethole . . . . 1 cc. to 3 quarts soln.
                (Worthley and Nicholas, 1937)

Molasses . . . . 100 cc.
Water . . . . 900 cc.
Nicotine sulfate . . . . 1 cc.
Pine tar oil . . . . 1 oz.
Valeric acid . . . . 0.5 cc.
                (Van Leeuwen, 1939)

Solid baits also have been employed and, while not consistently so effective as liquids, are reported to catch fewer flies, beetles, and other miscellaneous insects (Eyer, 1945). One of the best media is a mixture of gum arabic, cane sirup, sawdust, and water. The type of attractant chosen determines the ratio of ingredients. The mixture is added to a paper mold, e.g. a 4 oz. wax paper cup, which is suspended upside down over water in the same sort of canning jar used with liquid baits.

CONCENTRATION. Knowledge of the necessity of regulating the concentration of an attractant, gained through the use of olfactometers in the laboratory, can be applied with profit to operation in the field. A balance must be struck between the strength of attractant necessary to provide effective range and the optimum attractiveness. Usually a dilute solution or its equivalent is most effective. Dilution may be achieved by mixing the attractant with some neutral nonodorous diluent such as mineral oils. Langford, *et al.* (1943) reported that some attractants are equally or more effective with a diluent than in concentrated form. Addition of glycerin, for example, does not reduce the efficiency of Japanese beetle bait in proportion to the amount of glycerin added. A mixture of 8 parts of glycerin and ethyl alcohol to 2 parts of attractant caught only 10 per cent fewer beetles than the undiluted attractant. A mixture of 1 part pimento oil and 1 part ethyl alcohol possessed an attractive value approximately 10 per cent greater than pimento oil.

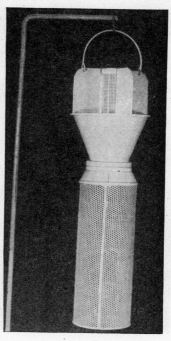

FIG. 42. One type of Japanese beetle trap showing the perforated cylinder, which contains the attractant, at the center of the baffles. (From Fleming *et al.*, *U.S. Dep. Agr. Circ.*, 1940.)

Instead of diluting the attractant to regulate aerial dilution one can achieve the same effect by controlled evaporation of the concentrated substance. Controlled and continued even evaporation may be attained through the agency of a wick, an added base, a plaster of Paris disk, a sponge, or a sack of fuller's earth. Generally, wicks are used in Japanese beetle traps where a bottle and wick dispenser insures a more or less constant rate of evaporation and a level of attractiveness (Fleming, Burgess, and Maines, 1940). A cotton wick encased in a woven cotton sheath the fibers of which extend lengthwise has been found most satisfactory. It should extend from the bottom of the bottle of geraniol and eugenol to 1 to 2 inches beyond the top. The bottle is placed in a perforated metal cylinder at the center of the baffle (Fig. 42) or mounted exposed in the center of the baffle with a cone for protection against the rain (Fig. 43).

In bucket type traps or bait pails it has been customary to mix the attractant with a base, an emulsifier, or both to control evaporation. In codling moth traps Bobb, Woodside, and Jefferson used a base or carrier solution of stock sirup (trade names White Mule and Karo) to which was added the attractant emulsified at the rate of 14 oz. to 3 oz. of bentonite to $\frac{1}{2}$ gal. of water. Eyer and Medler (1940) emulsified their attractant at the rate of 1 cc. or 1 g. to a 99 cc. mixture of gum tragacanth (200 g.), glycerin (100 cc.), and water (to make 2 l.) (trade name Emulsarome).

FIG. 43. Japanese beetle trap in which the bottle containing the attractant is protected from the rain by a small inverted cone. (From Fleming *et al.*, *U.S. Dep. Agr. Circ.*, 1940.)

Bait, either emulsified or straight, may be added to stock sirup as stated above or in an evaporation cup floating on standard sirup bait. The latter procedure is usually employed in testing the quality of new experimental attractants.

Purely for experimental purposes an atomizer or blower provides a highly satisfactory method of dispensing attractants when large quantities are available. In this manner dilution, evaporation, and dispersal are facilitated. The fumes, so-called, will attract when the identical material dispensed some other way may be practically neutral. Williams (1943) reported powerful attraction of kelp flies by trichlorethylene when the chemical was forced through a blower. Dispensed with a wick in a standard Japanese beetle trap the same chemical attracted no flies at all (Dethier, unpublished). Likewise, the powerful attraction of automobile paint for the palmetto weevil as reported by Bare (1929) occurred in the presence of clouds of fumes produced by an atomizer.

Materials which absorb odor in the process of enfleurage likewise emanate the same odor evenly over a long period of time. Their use as agencies for dispensing attractants has not yet been tried. Whether or not a high enough concentration could be maintained is problematical.

ORIENTAL FRUIT MOTH. Several baits for *Grapholitha molesta*, consisting of Karo, sucrose, or sorghum sirup as a base for fermentation by airborne bacteria and fungi plus oil of anise or terpinyl acetate and lignum pitch as a base have been developed (Bobb, 1938). Generally speaking, safrol, terpinyl acetate, oleic acid, linalool, and linolic acid are satisfactory attractants. Usually bait pails are used, the procedure being similar to that followed against codling moths.

TOBACCO HORNWORM. Attractants for use in traps against *P. sexta* have but recently been developed (Morgan and Crumb, 1928; Morgan and Lyon, 1928; Gilmore and Milam, 1933; Scott and Milam, 1943). It is reported that more adults are attracted to artificial flowers scented with amyl acetate than to the preferred food plants. Also highly attractive are benzyl benzoate, isoamyl benzoate, and isoamyl salicylate.

FRUIT FLIES. The use of so-called protein lures for control of fruit flies belonging to the genera *Dacus*, *Chaetodacus*, *Ceratitis*, *Anastrepha*, and *Rhagoletis* is being rapidly developed in this country. The basic attractant is a low concentration of ammonia produced by the hydrolysis of protein by alkali. Casein, gelatin, yeast, blood, egg white, wheat shorts, glycine, *dl*-alanine, or *l*-cystine in the presence of sodium hydroxide are attractive (McPhail, 1939). A bait consisting of 2 per cent glycine (tech.) plus 3 per cent commercial sodium hydroxide has been found to be effective against the walnut husk fly (Boyce and Bartlett, 1941). Against the apple maggot fly a lure composed of either powdered egg albumin, a peptone preparation, or dried yeast powder and alkali has given good results (Dean, 1941). Against the same insect a lure composed of commercial ammonia and ammonium sulfate plus soap to control evaporation is said to be superior to glycine (i.e. ammonia produced from glycine) (Hodson, 1943). In Europe, lures which contained ammonium phosphate have been used against the olive fruit fly (Bua, 1933, 1938; Bohorquez, 1940).

FIG. 44. A Japanese beetle trap with part of the cylinder wall removed to illustrate the manner in which currents of air pass through. (From Rex, *New Jersey Dep. Agr. Circ.*, 1933.)

MOSQUITOES. Thus far only one chemical compound has been reported as attractive to mosquitoes in the field. It is carbon dioxide (see p. 132). The use of carbon dioxide as an attractant in traps was developed by New Jersey workers (Headlee, 1934; 1941). A 3-lb. block of dry ice (carbon dioxide) is placed in a standard New Jersey light trap. As the carbon dioxide sublimes it is dispersed. The use of CO_2 as an adjunct to the New Jersey trap is still in the experimental stage since the efficiency of the trap itself is still under investigation (Huffaker and Back, 1943). Moreover, recent work has cast some doubt on the effectiveness of CO_2 as an attractant. A human being sleeping in a small hut to which mosquitoes can gain access but from which they cannot escape still constitutes the best bait.

MISCELLANEOUS. The following baits have been used with some success to trap the insects indicated. Vinegar or fermenting fruit sirup attracts *Ephestia kühniella* Zell., *E. elutella* Hb., *E. cautella* Wlk., *Plodia interpunctella* Hb., and *Corcyra cephalonica* Staint. (Noyes, 1930). Fish oil and similar products lure clothes moths, carpet beetles, etc. (Wilson, 1940). Liver-baited traps catch eye gnats. Vinegar and molasses in cone traps catch the alfalfa pest, *Caenurgia erechtea* (Cram.) (Smith, 1924). The roach, *Blatta orientalis*, is attracted to some essential oils (Cole, 1932). In thrip traps benzaldehyde, cinnamaldehyde, and anisaldehyde are effective attractants (Howlett, 1914). No outstandingly effective attractant has been found for use against houseflies. Sundry fermenting mixtures still are most useful.

Traps. A trap is merely a device for immediately killing or retaining for later destruction insects attracted by bait or light and insects that blunder in by chance. Different types must be developed for different species of insects. Representative types embodying the use of baits or attractants are described below. For descriptions of other bait traps and light, wind, and mechanical traps the reader is referred to Peterson's (1934) manuals.

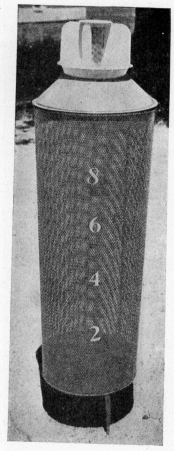

FIG. 45. A large Japanese beetle trap with a capacity of 40 quarts. (From Rex, *New Jersey Dep. Agr. Circ.*, 1933.)

METAL, WOODEN, AND GLASS TRAPS. The simplest type is the

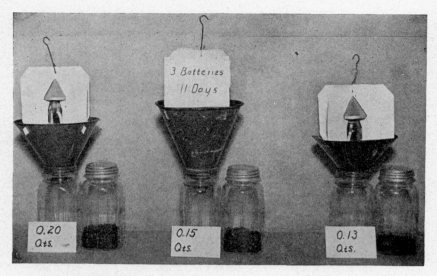

FIG. 46. The angle of funnel pitch is one factor which affects the efficiency of Japanese beetle traps. From left to right the funnel pitch is 60°, 45°, and 80°. (From Langford *et al.*, *J. Econ. Entomol.*, 1940.)

so-called bait pail with or without an evaporation cup. When an evaporation cup is incorporated, insects are caught and held by the solution on which the cup is floating. Without an evaporation cup insects are simply trapped in the bait where they drown. Soap or some other material may be added to assist in preventing the escape of the insects. The addition of baffles to knock insects into the solution may increase the efficiency of a trap as far as some species are concerned and lessen it as far as others are concerned. The latter effect was shown to be true in the case of 12-in. bait pans operated for codling moths by Worthley and Nicholas (1937).

More complex metal traps are those primarily intended for use with pure attractants dispensed with a wick or, in the case of mosquito traps, by sublimation. The Japanese beetle trap has gone through many modifications all directed toward cutting cost and increasing efficiency. Several models now in use (Figs. 44–48) are fundamentally similar, differing materially only in cost and efficiency. They consist of a four-winged baffle mounted on top of a funnel, a device for holding the bait dispenser, and a receptacle for holding the captured beetles. The present model is the result of exhaustive tests to determine the most satisfactory design of all

FIG. 47. The diameter of the aperture at the bottom of trap funnels also affects efficiency. The extent of variation found in traps is shown here. (From Langford et al., J. Econ. Entomol., 1940.)

FIG. 48. Types of containers used in Maryland for Japanese beetle traps. The two on the upper right are the most satisfactory. (From Langford et al., J. Econ. Entomol., 1940.)

parts (Langford, Cory, and Whittington, 1940; Fleming, Burgess and Maines, 1940). Considered best is a four-winged baffle extending 4 in. above the top of the funnel and 2 to 4 in. into the funnel, a funnel with a 60° pitch and a ¾ inch aperture at the bottom, a wick and bottle dispenser holder cut into the central portion of the baffle and possessing a small inverted cone to afford protection from the rain, and a well-ventilated elongate beetle container.

Though primarily a light trap, the New Jersey trap for mosquitoes has been adapted to dispense carbon dioxide (Fig. 50).

Trap design for tobacco hornworm control is still in the initial stages (Fig. 51), but Scott and Milam reported considerable success with a trap designated as the short-funnel model. It consists of a wood frame 36 in. long, 24 in. wide, and 34 in. high, covered on top and sides with wire screening and equipped at each end with a screenwire entrance funnel extending inwardly about 10 in. A hardware cloth baffle hung between the two funnels prevents moths from flying directly through the trap. Within the trap is placed a feeder, a green can in the top of which are fitted three small white tin funnels simulating a cluster of jimson blossoms, and a vial of isoamyl salicylate with a wick or a sack of fuller's earth. Moths are attracted initially by the odor and then visually by the white funnels. The feeder may be used to lure moths into a trap or it may be filled with a liquid poisoned bait containing 10 per cent of sugar with 5 per cent of tartar emetic or 0.04 per cent of rotenone.

FIG. 49. The Maryland 1940 Japanese beetle trap. (From Langford *et al.*, *J. Econ. Entomol.*, 1940.)

In designing traps to catch pomace flies the most formidable obstacle to overcome is the ease with which these small insects readily escape by crawling away. Of several traps developed Reed's (1938) was the first to prove eminently effective for small-scale trapping indoors. It is a piece of glass tubing 8½ cm. long and 28 mm. in internal diameter, stoppered at one end and funneled at the other (Fig. 52). The funnel opening is 2.1 mm. in diameter. Bait is placed in a glass dish or thimble and set on the cork. For large-scale

Fig. 50. A New Jersey mosquito trap baited with dry ice. (Courtesy, U.S. Department of Agriculture, Bureau of Entomology and Plant Quarantine.)

trapping in the field the standard trap is the glass model illustrated in Fig. 53. Recently, Hodson (unpublished) perfected a simple, cheap, effective trap for apple maggot flies. It consists of a cardboard cottage-cheese container with a perforated false bottom beneath which is placed ammonium carbonate. The insides of the container are smeared with an adhesive, and the trap is hung in orchards in an inverted position. It has proved very successful as a means of determining spray times.

Housefly traps (Fig. 55) properly operated constitute an effective

FIG. 51. An improved trap for capturing tobacco hornworm moths. (Courtesy, U.S. Department of Agriculture, Bureau of Entomology and Plant Quarantine.)

adjunct to standard methods of control. Innumerable models have been designed, all operating in the same manner. Flies attracted to molasses, fruit, milk, meat, or fermenting baits placed in shallow pans beneath the traps fly upward upon completion of feeding. Passing through a small opening they enter the screened enclosure from which they are unable to escape. Blowflies as well as houseflies are constant visitors. Conical traps are reportedly the most efficient type.

Traps to catch thrips are merely adaptations of the common bait pail. As simple an arrangement as a small bowl containing ½ pt. of water plus 2 cc. of benzaldehyde, cinnamylaldehyde, or anisaldehyde will catch large numbers of thrips.

Improvised traps baited with liver and water and placed adjacent to schools have aided materially in reducing the incidence of purulent conjunctivitis among school children by catching the vector,

Hippelates. Bait is placed at the bottom of a can which is equipped with openings near the base and a glass jar attached to the side

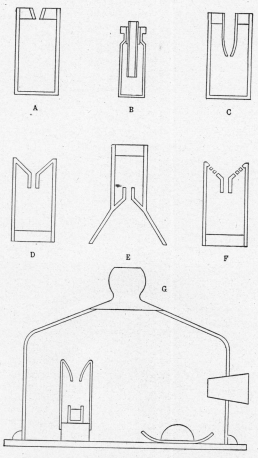

Fig. 52. (A to F) Different types of traps employed to catch *Drosophila*. (G) Reed's trap and experimental set-up to test the reactions of *Drosophila* to odorous substances. The odor solution is contained in a small vessel within the trap. Humidity is regulated by water-soaked cotton in a watch glass. (Redrawn after Reed, *Physiol. Zool.*, 1938.)

near the top (Fig. 56). Eye gnats attracted by the bait enter the can and upon completion of feeding fly up into the jar where they die.

FIG. 53. Standard glass trap for field trapping of fruit and pomace flies. Bait is placed in the circumferential trough of the base and flies enter through the basal opening.

FIG. 54. Trap baited with ammonium carbonate to capture apple maggot flies (*Rhagoletis pomonella* Cress.) (Courtesy, A. C. Hodson.)

ELECTROCUTING TRAPS. Electrocuting traps have been designed for use with baits, but the initial cost plust cost of operation is prohibitive in view of efficiency. Others are modeled for use against doors and windows to which flies are attracted by odors from within. All carry from 3,500 to 4,000 volts with low amperage.

LOG TRAPS. To date the only successful method of trapping bark beetles of the family Scolytidae consists of felling elm logs and de-

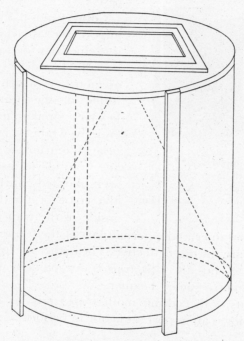

FIG. 55. Diagram of one type of standard housefly trap. This is a cylindrical wire screen container with a central screen cone.

stroying the logs after they have become thoroughly infested with beetles. In essence this is similar to planting trap crops (crops which are sown prior to the main crop, attract crop pests, and are destroyed with these pests). Elms treated and killed by the injection of ethyl alcohol or sodium chlorate are exceptionally attractive to certain species of bark beetles (cf. p. 98).

STICKY BANDS. Attempts to increase the efficiency of sticky bands by the addition of attractants so that insects will seek out the bands rather than blunder into them have been but partially successful. Among attractants that could be added to sticky bands for control of the pear midge, *Contarinia pyrivora*, nitrobenzene, oil of cloves, and oil of fennel show promise (Staniland and Walton, 1930). In factories, warehouses, etc., sticky materials to which have been added attractants are successful in preventing extensive infestation by flying grain, cocoa, and fabric pests.

FIG. 56. An improvised eye gnat trap baited with liver. (Courtesy, U.S. Department of Agriculture, Bureau of Entomology and Plant Quarantine.)

Color. Color has lately assumed importance as a factor affecting the efficiency of some traps, especially Japanese beetle traps. A green trap in which the baffle and inside of the funnel are white behaves similar to one painted entirely white. When these parts of a white trap are painted green, the trap performs as though it were painted entirely green Both the color of the baffle and the inside of the trap are important, that of the container and external part of the funnel unimportant. Yellow traps are the most effective. The higher the luster the greater the efficiency. Thus the best trap is one painted with a primary yellow "automotive-type" lacquer.

Experiments with codling moth traps thus far have not indicated that color is an important factor in performance (Bobb, *et al.*, 1939).

Baker, *et al.*, (1944) reported that color is of some importance in trapping the Mexican fruit fly. More than twice as many flies are caught with yellow as with any other color except dark green.

Position and Location of Traps. Various external factors greatly affect the efficiency of traps. These include the season, activity of the insect, weather, and the location of the trap with reference to source of infestation, the presence of preferred food plants, and buildings or other obstructions. Since insects follow odors upwind, they are attracted from the leeward. It is essential that wind carrying odors be free to blow for a great distance. Obstructions such as buildings, trees, hills, etc., which obstruct or impede air movement reduce the effective range of operation. Japanese beetle traps may attract from a distance of 300 to 500 yards if placed on the windward side of an open field. Best results obtain when a trap is suspended in an open sunny position to the windward of plants subject to attack and not closer than 10 to 25 ft. of plants on the lee-

FIG. 57. Ideal location in an apple tree for a codling moth bait pail. (From Bobb *et al.*, *Virginia Agr. Exp. Sta. Bull.*, 1939.)

ward. For the protection of orchards best results obtain when traps are placed on the windward side at a distance of 10 to 25 ft. rather than throughout the orchard. Recently, it has been learned that traps operated 5 to 10 ft. from preferred food plants catch from 2 to 2½ times as many beetles as traps located 10 ft. away. Variation in catch in different adjacent areas is due to the kind and abundance of host plants (Whittington, *et al.*, 1942). This phenomenon is quite understandable when one remembers that the odor from a trap is competing with the odors of natural surroundings. Proper vertical location is important also. A height of $3\frac{1}{2}$ ft. is better than $4\frac{1}{2}$ or $5\frac{1}{2}$ ft.

Studies over a period of years have shown that codling moth traps should be suspended near the tops of trees (Bobb, *et al.*, 1939; Spuler, 1927; Yothers, 1927) where codling moths do most of their flying. Moreover, traps are most efficient when placed in trees with open centers (Fig. 57) where they do not bump branches and where moths can dart back and forth.

Sex Ratio. Both sexes of a species are not always caught in traps in equal numbers. This phenomenon may be due to a difference in response to a given attractant (cf. p. 19) or to a difference in activity behavior. Thus mosquito traps baited with carbon dioxide are said to catch chiefly females because carbon dioxide is repellent to males. On the other hand, apparent sexual differences in response as evinced by trap catches frequently are due to differences in behavior. Trap catches of codling and oriental fruit moths in which females predominated have been reported by many workers (Peterson, 1925; Yothers, 1927; Steiner, 1929). In one case 50 to 60 per cent of the catch were females of which 95 per cent were gravid. Supler reported that 60 per cent of 17,429 moths caught were females. Bobb, *et al.* reported a female catch of 59.8 per cent. That the greater percentage of females caught results from the fact that females remain close to trees and are less active is evinced by the fact that rotary nets sampling a higher aerial stratum catch predominantly males (Alexander and Carlson, 1943). In other instances sex ratios different from the expected are not so easily explained. Whereas the ratio of sexes in reared populations of the raisin moth, *Ephestia figulilella* Greg., and the grape leaf roller, *Desmia funeralis* (Hbn.), was approximately 1, baited traps caught less than 50 per cent females for the raisin moth and more than 50 per cent for the grape leaf roller (Barnes, 1943). Also unexplained is the predominance of males of the mango fruit fly, *Chaetodacus dorsalis* Hendel., caught by a bait composed of soap solution, eugenol, and lemon grass oil (Miwa and Moriyama, 1940). In view of McPhail's recent work (cf. p. 116) it may be that the flies actually are responding to fatty acids in the soap and that in this case these are sex attractants. Observed differences between the numbers of male and female Mexican fruit flies caught in baited traps are thought by Baker, *et al.* to be due to the influence of oviposition responses in the females.

Reliability of Trapping as a Means of Sampling Population. As a means of locating new infestations and of determining emergence time, proper time for spraying, and peak breeding, traps are highly reliable. Without a doubt they have been more successful in this rôle than in any other. If a trap attracts a species at all, it can be used to good advantage in any of the above-mentioned capacities.

When a trap which is nonspecific is set up to sample mixed populations, it cannot present a true representation of comparative species density. Mosquito traps are a case in point. Since different species of mosquitoes react differently to the same stimulus, the results of catches of each are not comparable unless corrections can be made to offset innate differences in activity. An intensive study of the efficiency of the New Jersey mosquito trap in this respect has been made by Huffaker and Back (1943). They concluded that none of the current trapping methods could be depended upon for an adequate nonselective analysis of a heterogeneous mosquito population.

Efficiency of Trapping as a Means of Control. Opinion as to the value of traps as agents of control has, generally speaking, been unfavorable. Considerable evidence has been amassed to show that rather than reducing population density in a given area, traps caused an increase in local population which they could not cope with adequately. Immigration from adjacent areas presumably accounted for the increase. Van Leeuwen (1939) concluded that such action occurred when codling moth traps were set out. Alexander and Carlson (1943), on the other hand, presented new evidence that traps in scattered trees do not attract moths from any great distance. They compared catches by bait traps and rotary net. The latter is a mechanical trap, an electrically driven rotary net, developed by Chamberlin and Lawson (1940) for sampling aerial insect populations (Fig. 58). Graphs comparing bait and net catches (Fig. 59) show beautiful correspondence between the two methods. Of 350 marked moths that were released, 23 were retaken in nets and 46 in traps. The conclusion that baits do not encourage marked immigration is based on the fact that nets recaptured nearly as many as did traps.

The value of codling moth traps as a control measure is, however, strongly argued by the experiments of Yothers (1927) and
Bobb, *et al.* (1939). In the first instance, in a baited block of 42
trees in Washington, counts of harvested fruit showed from 12 to
16 per cent more clean fruit than the same varieties similarly

Fig. 58. Rotary nets for sampling the codling
moth population at different heights in an
orchard. (Courtesy, U.S. Department of Agriculture, Bureau of Entomology and Plant
Quarantine.)

sprayed but not protected by bait traps. In the second instance all
trees in an orchard were banded (chemically treated). In one of
these sections of the orchard a trap was placed in each tree. An
examination of these bands in the fall revealed that approximately
six times as many larvae were captured in bands in unbaited blocks
of trees as in the baited block. The evidence that trapping as an
adjunct to spraying is a worthwhile means of control is significant.
It is apparent that (1) many females are caught, (2) as much as

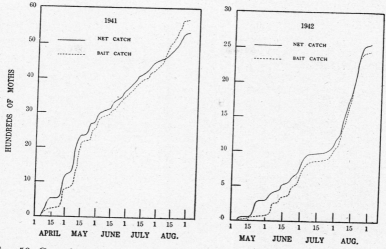

Fig. 59. Cumulative numbers of codling moths caught in a rotary net and the average catch of two molasses bait traps in 1941 and 1942, Yakima, Wash. (Redrawn after Alexander and Carlson, *J. Econ. Entomol.*, 1943.)

95 per cent may be gravid, (3) less than 5 per cent (Bobb, 1938) may have deposited their full quota of eggs, (4) the number of larvae is correspondingly reduced, (5) there is no significant migration into the orchard, and (6) a higher percentage of clean fruit results.

Control of the Japanese beetle in densely infested areas by means of traps is apparently less decisive. Erection of as many as 100 traps to the square mile of heavily infested area causes a minor reduction in local population. There is evidence that a sufficient number of immigrants may be attracted to increase the density. Over a period of years, however, sufficiently large-scale trapping may reduce or retard general infestation.

Ammonium and protein lures employed in the control of fruit flies apparently do not suffer the fault of causing immigration (Hodson, 1943).

REFERENCES

Alexander, C. C., and F. W. Carlson: A comparison of codling moth captures by bait trap and rotary net, *J. Econ. Entomol.*, 36(4), 637–638 (1943).

Baker, A. C., W. E. Stone, C. C. Plummer, and M. McPhail: A review of studies on the Mexican fruitfly and related Mexican species, *U.S. Dep. Agr., Misc. Pub.* 531, 155 pp. (1944).

Bare, C. O.: *Rhynchophorus cruentatus* Fab., the palmetto weevil, attracted to automobile paint, *J. Econ. Entomol.*, 22(6), 986 (1929).

Barnes, D. F.: Influence of moth-trapping methods on the proportion of females in catches, *Ibid.*, **36**(1), 119–120 (1943).

Bobb, M. L.: The use of bait traps in codling moth control, *Virginia Fruit*, **26**(4), 20–24 (1938).

——, A. M. Woodside, and R. N. Jefferson: Baits and bait traps in codling moth control, *Virginia Agr. Exp. Sta., Bull.* 320, 1–19 (1939).

Bohorquez, R.: Experiencias de lucha contra la mosca del olivo (*Dacus oleae* Gmel.) por medio de sustancias atractivas, *Bol. pat. veg. ent. agr.*, **9**, 188–204 (1940).

Boyce, A. M., and B. R. Bartlett: Lures for the walnut husk fly, *J. Econ. Entomol.*, **34**(2), 318 (1941).

Brown, H. E.: The use of attractants in codling moth sprays, *Proc. Missouri State Hort. Soc.*, **1941–1942**, 78–80.

Bua, G.: Experimenti dell 1933 con sostanze attrative per la mosca delle olive, *Ann. ist. super. agrar. Portici*, **6**(3), 125–145 (1933).

——: Serie di esperimenti con sostanze attrative per la mosca delle olive, *L'Olivicoltore*, **15**, 1–19 (1938).

Burdette, R. C.: Attraction of certain insects to spray baits, *J. Econ. Entomol.*, **25**(2), 343–346 (1932).

Chamberlin, J. C., and F. R. Lawson: A mechanical trap for the sampling of aerial insect populations, *U.S. Dep. Agr., Bur. Entomol. & Plant Quarantine*, ET-163, 1940, 6 pp.

Chisholm, R. D., and L. Koblitsky: Variation in composition and volatility of Japanese beetle attractants during evaporation, *J. Econ. Entomol.*, **38**(4), 467–470 (1945).

Cole, A. C.: The olfactory responses of the cockroach (*Blatta orientalis*) to the more important essential oils and a control measure formulated from the results, *Ibid.*, **25**(4), 902–905 (1932).

Dean, R. W.: Attraction of *Rhagoletis pomonella* adults to protein baits, *Ibid.*, **34**(1), 123 (1941).

Dethier, V. G.: Chemical factors determining the choice of food plants by Papilio larvae, *Am. Naturalist*, **75**, 61–73 (1941).

——, and F. H. Whitley: Population studies of Florida mosquitoes, *J. Econ. Entomol.*, **37**(4), 480–484 (1944).

Ditman, L. P., J. P. Secrest, and E. N. Cory: Studies on corn ear worm control, *University of Maryland Agr. Exp. Sta., Bull.* 439, 205–223 (1941).

Eyer, J. R.: Solid baits for codling moth, *J. Econ. Entomol.*, **38**, 344–346 (1945).

——, and J. T. Medler: Attractiveness to codling moth of substances related to those elaborated by heterofermentative bacteria in baits, *Ibid.*, **33**(6), 933–940 (1940).

Fleming, W. E., E. D. Burgess, and W. W. Maines: The use of traps against the Japanese beetle, *U.S. Dep. Agr., Circ.* 594, 1–11 (1940).

——, and R. D. Chisholm: Anethole and pimenta leaf oil as attractants for the Japanese beetle, *J. Econ. Entomol.*, **37**(1), 116 (1944).

Gaddum, E. W.: Some observations on jassid at the Kenya coast, *Empire J. Exp. Agr.*, **10**(39), 133–145 (1942).

Gilmore, J. U., and J. Milam: Tartar emetic as a poison for the tobacco hornworm moths, a preliminary report, *J. Econ. Entomol.*, **26**(1), 227–233 (1933).

Headlee, T. J.: Mosquito work in New Jersey for the year 1933, *Proc. New Jersey Mosq. Exterm. Assoc.*, **21**, 8–37 (1934).

——: New Jersey mosquito problems. *Proc. 28th Ann. Meeting, New Jersey Mosq. Exterm. Assoc.*, **1941**, pp. 7–12.

Heriot, A. D.: How does lead arsenate prevent the young codling moth larva from injuring the fruit? *Proc. Entomol. Soc. British Columbia*, **40**, 3–8 (1943).

Hodson, A. C.: Lures attractive to the apple maggot, *J. Econ. Entomol.*, **36**(4), 545–548 (1943).

Howlett, F. M.: A trap for thrips, *J. Econ. Biol.*, **9**(1), 21–23 (1914).

Huffaker, C. B., and R. C. Back: A study of methods of sampling mosquito populations, *J. Econ. Entomol.*, **36**(4), 561–569 (1943).

Hutner, S. H., H. M. Kaplan, and E. V. Enzmann: Chemicals attracting *Drosophila*, *Am. Naturalist*, **71**, 575–581 (1937).

Jones, H. A., and H. L. Haller: Properties of two samples of commercial geraniol used in Japanese beetle traps, *J. Econ. Entomol.*, **33**(2), 327–329 (1940).

——, and ——: Bieugenol in a commercial geraniol, *J. Am. Chem. Soc.*, **62**, 2558 (1940).

——, and ——: Composition of geraniol for Japanese beetle bait, *News Ed.* (*Am. Chem. Soc.*), **19**, 683–685 (1941).

Koblitsky, L., and R. D. Chisholm: Standard method for the distillation of geraniol, *U.S. Dep. Agr., Bur. Entomol. & Plant Quarantine, Circ.* E-511, 1940.

Langford, G. S., and E. N. Cory: Japanese beetle attractants with special reference to caproic acid and phenyl ethyl butyrate, *J. Econ. Entomol.*, **39**(2), 245–247 (1946).

——, ——, and F. B. Whittington: Inexpensive Japanese beetle traps, *Ibid.*, **33**(2), 309–316 (1940).

——, M. H. Muma, and E. N. Cory: Attractiveness of certain plant constituents to the Japanese beetle, *Ibid.*, **36**(2), 248–252 (1943).

Loeffler, E. S., and W. M. Hoskins: Evaluation of blow-fly repellents, "Laboratory Procedures in Studies of the Chemical Control of Insects," Washington, D. C., Am. Assoc. Adv. Sci., 1943, p. 173.

McPhail, M.: Protein lures for fruitflies, *J. Econ. Entomol.*, **32**(6), 758–761 (1939).

Miwa, Y., and T. Moriyama: Experiments with baits for the mango fruit fly (*Chaetodacus dorsalis* Hendel.), *Formosan Agr. Rev.*, **36**, 685–716 (1940).

Morgan, A. C., and S. E. Crumb: Notes on the chemotropic responses of certain insects, *J. Econ. Entomol.*, **21**(6), 913–920 (1928).

——, and S. C. Lyon: Notes on amyl salicylate as an attrahent to the tobacco hornworm moth, *Ibid.*, **21**(1), 189–191 (1928).

Munger, F.: Reactions of the citrus thrips to sugar in poisoned baits, *Ibid.*, **35**(1), 51–53 (1942).

Noyes, W. M.: Moth pests in cocoa and confectionery, *Bull. Entomol. Research*, **21**(1), 77–121 (1930).

Peterson, A.: A bait which attracts the oriental peach moth (*Laspeyresia molesta* Busck), *J. Econ. Entomol.*, **18**(1), 181–190 (1925).

——: "A Manual of Entomological Equipment and Methods," Part 1, 3d ed. Edwards Brothers, Ann Arbor, Mich.; Part 2, John S. Swift Co., St. Louis, 1934.

Reed, M. R.: The olfactory responses of *Drosophila melanogaster* Meigen to the products of fermenting banana, *Physiol. Zool.*, **11**(3), 317–325 (1938).

Rex, E. G.: Instructions for maintaining Japanese beetle traps, *New Jersey Dep. Agr., Circ.* 235, 1–18 (1933).

Richmond, E. A.: Olfactory response of the Japanese beetle (*Popillia japonica* Newman), *Proc. Entomol. Soc. Washington*, **29**(2), 36–44 (1927).

Ripley, L. B., and G. A. Hepburn: Further studies on the olfactory reactions of the Natal fruit-fly, *Pterandrus rosa* Ksh., *Union S. Africa, Dep. Agr. Mem.* 7, 24–81 (1931).

Rudolfs, W.: Effect of chemicals upon the behavior of mosquitoes, *New Jersey Agr. Exp. Sta., Bull.* 496, 1–24 (1930).

Schreiber, A. F.: Deterrents to insects (in Russian), *Horticulturist* (Rostov-on-Don), **15**(1), 50–51. *Rev. Applied Entomol., A* **4**, 161 (1916).

Scott, L. B., and J. Milam: Baits and traps for the control of tobacco and tomato hornworms, *U.S. Dep. Agr., Bur. Entomol. Plant Quarantine*, E-578, 1942.

——, and ——: Isoamyl salicylate as an attractant for hornworm moths, *J. Econ. Entomol.*, **36**(5), 712–715 (1943).

Siegler, E. H.: Laboratory studies of codling moth larval attractants, *Ibid.*, **33**(2), 342–345 (1940).

——, and H. A. Jones: Extracts of apple peels as adjuvants to lead arsenate against codling moth larvae, *Ibid.*, **35**(2), 225–226 (1942).

Smith, R. C.: *Caenurgia erechtea* (Cram) (Noctuidae) as an alfalfa pest in Kansas, *Ibid.*, **17**, 312–320 (1924).

Spuler, A.: Codling moth traps, *Washington Agr. Exp. Sta. Bull.*, 214, 1–12 (1927).

Stabler, R. M.: New Jersey light-trap versus human bait as a mosquito sampler, *Entomol. News*, **56**(4), 93–99 (1945).

Staniland, L. N., and C. L. Walton: Progress report. Experiments on the control of pear midge (*Contarinia pyrivora*). *Ann. Rep. Agr. Hort. Research Sta., Univ. Bristol*, **1929**, 124–129 (1930).

Steiner, L. F.: Codling moth bait trap studies, *J. Econ. Entomol.*, **22**(4), 636–648 (1929).

U.S. Dep. Agr., Bureau of Entomology and Plant Quarantine: Insects in Relation to National Defense, Circ. 8, **1941**, 33 pp.

Van Leeuwen, E. R.: Investigations of baits attractive to the codling moth, *Proc. 31st Ann. Meet. Washington State Hort. Assoc.*, **1935**, 136–139.

——: Further contributions to a study of baits for the codling moth, *Proc. 35th Ann. Meet. Washington State Hort. Assoc.*, **1939**, 21–30.

——: Chemotropic tests of materials added to standard codling moth bait, *J. Econ. Entomol.*, **36**(3), 430–434 (1943).

——, and F. W. Metzger: Traps for the Japanese beetle, *U.S. Dep. Agr., Circ.* 130, 2–5 (1930).

Walton, R. R., and F. E. Whitehead: Tests of ingredients of grasshopper baits, *J. Econ. Entomol.*, **38**(4), 452–457 (1945).

Whittington, F. B., E. N. Cory, and G. S. Langford: The influence of host plants on the local distribution of Japanese beetles and on the effectiveness of traps, *Ibid.*, **35**(2), 163–164 (1942).

Williams, C. M.: Dry-cleaning fluid as an attractant for the kelp-fly. "Laboratory Procedures in Studies of the Chemical Control of Insects," Washington, D.C., Amer. Assoc. Adv. Sci., 1943, p. 174.

Wilson, H. F.: Lures and traps to control clothes moths and carpet beetles, *J. Econ. Entomol.*, **33**(4), 651–653 (1940).

Wingo, C. W., and H. E. Brown: Field studies of codling moth larvae attractants, *Ibid.*, **35**(2), 284–285 (1942).

Worthley, H. N.: Emergence cages and bait pails for timing codling moth sprays, *Penn. Agr. Exp. Sta. Bull.*, 277, 1–19 (1932).

——, and J. E. Nicholas: Tests with bait and light to trap codling moth, *J. Econ. Entomol.*, **30**(3), 417–422 (1937).

Yothers, M. A.: Summary of three years' tests of bait traps for capturing the codling moth, *Ibid.*, **20**(4), 567–575 (1927).

Chapter 8

Repellents

Definition and Action. Repellents are those substances which as stimuli elicit avoiding reactions from organisms. They fall, as do attractants, into two broad categories, physical and chemical. In these states they serve as warning stimuli. Whereas attractants are designed to lead an insect to its food, its mate, its oviposition site, repellents operate primarily in a protective capacity, warding off enemies or warning an insect of enemies, improper food, and inimical surroundings. In sharp contrast to attractants which are outstandingly specific, repellents are notably nonspecific in their action. Dangers are proverbially nonselective. Instances where one or more insects are not repelled by stimuli which ordinarily repel a majority of species usually indicate special adaptation on the part of the few.

In nature where various environmental and ecological factors are repellent we find that the most important physical factors are associated with surface phenomena and that the most important chemical factors are of plant origin. The former occur more generally but are more likely to be taken for granted.

Physical Repellents. Contact stimuli at surfaces inform an insect as it alights or progresses of the nature of the substratum or of objects with which it comes in contact. From the response elicited we may deduce that certain surfaces and contacts are repellent. Among plants the physical characters of the leaves, stems, seeds, etc., may determine or at least influence the acceptability of the area as food or oviposition site. Hairs, spines, pubescence, waxy, leathery, or fibrous cuticles, difference in water content, tissue texture, and leaf shape are a few such characters that may act as repellents. Dust, water in the form of bubbles or a film, and sticky materials like resins also may repel. Surface characters of insects themselves as for example hair, pile, spines, and variously designed armor protect their owners by repelling attackers. Emulsions such as bathe spittle insects serve a similar purpose.

201

From observations of the efficiency of these man has learned to fashion three to his use; dust, sticky materials, and the natural resistance characters of plants.

In that it interferes mechanically with walking, maintaining a foothold, and feeding, dust is inherently repellent. Particle size is extremely important. For example, Ripley, Petty and van Heerden (1936 and 1939) showed that the degree of repellence of cryolite to larvae of the wattle bagworm, *Acanthopsyche junodi* Heyl., is correlated not with the amount of dust per square centimeter of leaf area but with the particle size inasmuch as cryolite containing coarse particles was repellent until sieved through silk. Glass or cryolite particles more than 10 microns in diameter definitely repel normal larvae. It has been shown repeatedly that nontoxic dusts are repellent. Cucumber leaves dusted with wood ash are immune to feeding by *Ceratia similis;* those sprayed with water steeped in ash and filtered are acceptable, indicating the repellent action of the ash as a dust (Moriyama, 1939).

Much work has been done with repellent poison dusts, but it is outside the domain of this book. The reader is referred for further information to Shepard (1941) and Martin (1940). Suffice it to say that one reason why toxicity of a dust increases as particle size is reduced (to a certain point) is that repellency decreases with particle size and more of the dust is consumed.

Sticky materials have long been used to prevent the access of insects to a given area, as for example the crown of a tree. Bands treated with adhesive materials are applied to tree trunks to prevent the ascent of larvae of the gypsy and codling moth and the tent caterpillar. They also serve as traps. In Kenya, banding has been tried as a means of controlling mealy bugs indirectly by preventing the access of ants (James, 1930). Generally speaking, any sticky material is effective.

By recognizing that certain characters of plants—chemical, physical, and physiological, but especially the latter two—by nature of their repelling action render plants resistant to insect attack, geneticists have succeeded in producing resistant strains. Characters found to be important in this respect include, in addition to those already mentioned, color, small stomata, repellent chemical compounds, lignification, vigor, precocity, quick recovery from wounds, seasonal adaptation, absence of response to specific

FIG. 60. Field test of maize grown by the Instituto Fitotecnico de Santa Clara, Argentina, and exposed to attack by locusts. Notice the striking difference between susceptible and resistant varieties. (Photograph by S. Horovitz and A. H. Marchioni.)

stimuli, copious flow of sap, etc. (Wardle and Buckle, 1923; Folsom and Wardle, 1934; Painter, 1936, 1941, 1943; Sweetman, 1936). It has been shown, for example, that resistance to leaf hopper injury is due in part to rough hairy pubescence (Jewett, 1932; Johnson and Hollowell, 1935), that a disturbed water content renders the cotton plant more susceptible to attacks by sap-feeding insects (Mumford and Hey, 1930), that the different attractive values of various species of *Acacia* for bagworms may be due to difference in water content affecting the physical properties of foliage (Ripley, Petty and van Heerden, 1939), that pines are less susceptible to attack by small gypsy moth larvae because of the shape of the needles, that oviposition by *Bruchus quadrimaculatus* Fab. is made on the basis of seed coat surface (Larson, 1927), that oviposition restrictions of different races of the snowy tree cricket, *Oecanthus niveus* De Geer, are governed by the size of plant stems (Fulton, 1925).

Painter has examined the phenomena of plant resistance at

great length, and his classification of resistance factors is given in part below.

    1. *Evasion:* Transitory or pseudoresistance by avoidance.
        a. Time of maturity (genetic and ecological).
        b. Soil conditions (largely ecological).
        c. Water balance (largely ecological).
    2. *Resistance and tolerance* (largely genetic).
        a. Resistance mechanisms largely of plant origin.
            (1) Resistance to diseases carried by insects.
            (2) Passive resistance; resistance due to mechanical barriers such as hairs, etc.
            (3) Active resistance.
                (a) Repair and recovery.
                (b) Tolerance mechanisms.
                (c) Destruction of eggs and other stages by cell proliferation.
                (d) Habit of growth.
        b. Resistance mechanisms requiring an insect counterpart (largely genetic).
            (1) Oviposition responses.
                (a) To surfaces.
                (b) To odorous substances.
                (c) To habit of growth of plant.
            (2) Mechanisms related to feeding with the following results for insects:
                (a) Abnormal length of life.
                (b) High mortality.
                (c) Reduced size.
                (d) Reduced fecundity.
    3. *Immunity* (largely genetic—no true immunity known).

This outline serves to emphasize what has already been said regarding the importance of plant repellents in the establishment of a resistant variety. To develop such a variety plant breeders must conduct a long series of controlled experiments involving a clear understanding of the relations between the insect and the plant, especially attractive and repellent qualities influencing choice on the part of the insect and the inheritance of these characters in the plant. Painter cites examples to illustrate that plant genetic factors for insect resistance can be transferred interspecifically. A case in point is the transference of genes for resistance of wheat to Hessian fly from *Triticum durum* to *Triticum vulgare* by way of Marquillo spring wheat. With knowledge of this sort it is possible to produce resistant varieties by selective breeding or by hybridization. A third method is that of grafting to a resistant stock.

Naturally Occurring Chemical Repellents. Under natural conditions chemical repellents serve the purpose of (1) protecting insects from their enemies, (2) regulating certain aspects of behavior associated with reproduction, (3) influencing food preferences. In the first capacity they are substances elaborated in the bodies of many species in the so-called stink or repugnatorial glands. Hemiptera and Coleoptera lead the list as proud possessors of glands of this type chief among them being stink bugs (Pentatomidae), assassin bugs (Reduviidae), certain Coccinellidae, and the fireflies (Lampyridae). How effective the odors, stenches, and irritants so produced are in repelling enemies is open to question. They are effective to a progressively lesser degree against insects, lizards, and birds. In but few cases is the chemistry of these products known though it is suspected that in many cases the odor is elaborated from waste or poisonous products taken with the food plant. Larvae of *Melasoma populi* (Chrysomelidae) secrete salicylaldehyde (Garb, 1915); adults of the beetle *Aromia moschata*, an ester of salicylic acid (Hollande, 1909; Smirnow, 1911); larvae of the brassy willow beetle (*Phyllodecta vitellinae* L.), salicylaldehyde (Wain, 1943); *Papilio* larvae through their eversible pouches, probably some poisonous compound from the food plants Umbelliferae and Aristolochiaceae (Schulz, 1911; Wegener, 1923); the larvae of *Notodonta concinnula*, strong HCl (Denham, 1888); larvae of *Dicranura vinula*, 0.5 g. of formic acid said to be 40 per cent (Poulton, 1890); *Carabus*, butyric acid (Marchal, 1910); the bombardier beetle, *Brachinus*, a corrosive vapor, probably oxides of nitrogen, boiling at 8° to 10° C.; Paussid beetles, a vapor reportedly containing free iodine (Fredericq, 1910).

Regulation of behavior associated with reproduction is seen in the inhibition of copulation and oviposition by repellent body odors and secretions. Occasionally, females that have just been fertilized not only lose their attractiveness to males but actually exhibit repellence. More commonly repellent odors may inhibit oviposition. Females of *Trichogramma evanescens*, for example, after having oviposited in a host, secrete a substance which inhibits oviposition in the same host by other females (Flanders, 1944). Additional regulation of oviposition by way of influencing the selection of sites, animal or vegetable, operates in the same manner as that affecting the choice of food. This is best examined in phytophagous species.

In plants both volatile and nonvolatile compounds occur in wide variety. Obviously, the former tend to act as olfactory repellents and the latter as gustatory repellents. Unlike attractants, chemical repellents may be effective in either of two forms because there are at least two classes of chemical stimuli that may repel—tastes and odors. It has been explained that tastes cannot function as attractants because they lack the quality of being able to attract an insect from one locality to another although in the case of aquatic insects this restriction does not hold. On the other hand they can, like odors, cause an insect to depart from a given area. Against phytophagous insects, therefore, usually the first repellents which are operative are the essential oils, gums, and resins. The second line of defense is made up of the nonodorous taste substances especially salts, acids, alkaloids, tannins, etc. For phytophagous species the taste of a plant is not so important in determining whether it will be attacked as to what degree it will be consumed (cf. p. 42). It has been shown that acid juices (e.g., malic acid) are a cause of avoidance of certain varieties of cherry by the cherry fruit fly in Europe; that wattle bagworms are repelled by most salts (magnesium sulfate, sodium chloride) and certain acids (tartaric, oxalic, tannic) and that some species of African grasshoppers are repelled by sodium bicarbonate (Lea and Nolte, 1941). These are a representative few of the gustatory repellents that may be operative. Since observers working with experimental repellents have not always been at pains to differentiate between the two classes of repellents, many substances listed in the following pages may really be gustatory rather than olfactory.

Natural olfactory repellents are responsible for the immunity of many plants. Two examples may be cited here; others are listed further on under appropriate headings. The nun moth, *Liparis monacha* L., is alleged to shun certain pines because of the presence in their needles of approximately 5 per cent turpentine not present in the needles of less immune pines (Sweetman, 1936). Teak and cypress pines are immune to termite attack on account of a sesqui-terpene (Oshima, 1919) (see also under termites).

Olfactory Repellents. It may be said that most odors are repellent if their concentration is great enough. Repellency and the degree of repellency are thus bound up with concentration. The perplexing question of whether repellent gaseous mixtures act as repellent odors through the medium of localized olfactory receptors or as irritants through the medium of the more generalized common chemical sense has not yet been satisfactorily answered. It is quite true that some substances behave as attractants at low concentra-tions and are indubitably olfactory stimuli, but when they repel at higher concentrations are they affecting the same end organs or a different set? Which receptors do olfactory repellents ordinarily stimulate? These and other questions remain.

Many repellents also are effective insecticides. Some possess very efficient knockdown qualities. The odor of oil of cloves, for example kills ants very quickly. Camphor and naphthalene, on the other hand, act very slowly (Stäger, 1933). There is no reason or justifica-tion, however, for employing the terms insecticide and repellent loosely or in a manner suggesting association, for there are repel-lents lacking insecticidal properties, and many insecticides, notably DDT, possess no repellent qualities whatsoever.

In most instances a repellent must compete with an attractant. Thus while many are repellent to varied groups of insects when present alone, they are no longer repellent when an attractant also is present. Here they must be concentrated enough to mask or obscure the attractant in addition to being repellent *per se*. There are a few natural substances that are sufficiently strong in this respect to act alike with all insects. In other words, a given com-pound by itself in sufficient concentration may repel many random species of insects yet it would not repel species A when placed with a substance attractive to species A, or B when placed with a sub-stance attractive to species B. It may be said that repellents are nonspecific when used alone but when competing with attractants show a group specificity related to their ability to mask a given attractant. A statement to the effect that repellents generally are hydrocarbons would merely indicate that hydrocarbons can best compete against certain attractants. Too little is known of repellents to state that certain chemicals *per se* are best. Among synthetic materials certain groups may be said to be generally repellent since they produce best results when competing against attractive lures such as man, cattle, and other vertebrates.

In the final analysis many so-called repellents are not repellents at all. They are deodorizers that counteract the effect of attractive odors by (1) absorption, (2) adsorption, (3) chemical alteration of potentially odorous compounds, (4) masking of attractive odors, or (5) initial inhibition of attractive odors during formation. There has been little work with absorption that is of entomological value The moderate repellent effect of animal charcoal is probably due to its powers of adsorption. Examples of the alteration of odorous compounds are not uncommon but are of little practical value at the present time. Within limits, the odor of ammonia can be neutralized by the addition of hydrochloric acid to form the non-odorous ammonium chloride; certain obnoxious gases may be rendered odorless by oxygenation or chlorination; odorous un-saturated compounds may be rendered odorless by hydrogenation; attractants in potato leaves are said to be destroyed by repellents formed through natural fermentation of the leaf juices (cf. p. 70). Most attractive compounds derived from fermentation and decom-position can be eliminated as odors simply by inhibiting the chemical process involved. A number of compounds may act as agents of this inhibitory process (see Chapter 4). None of the com-pounds falling into the above mentioned categories are truly repellents unless they possess a repellent action of their own.

Screening and Evaluating Candidate Repellents. Standard procedures for testing repellents consist of (1) determining the repelling power of the pure substance in the laboratory and (2) determining repellency by the ability of a test material to counter-act an artificial attractant or one to which the animal is normally subject in nature (cf p 174). The most accurate laboratory method of determining absolute repellency in terms of threshold is by means of an olfactometer as previously described. The second method is more commonly employed.

The material to be tested is applied to a natural food, to a natural oviposition site, or to an attractive bait. Its ability to prevent feeding or oviposition, as the case may be, is a measure of its effectiveness as a repellent. Different modifications of this procedure are neces-sary with different insects. Two generalized methods are given below by way of example, and others are described later under the name of the species being tested.

With phytophagous species information concerning exact dosage, while desirable, is not so necessary as in testing insecticides. Nevertheless, the same methods of material application and measurement of leaf area consumed may be employed. Experimental repellents may be applied by means of specially designed atomizers or air brushes operating as atomizers. Entire plants which are to be treated may be placed on a turntable to insure even coverage while individual leaves may be placed at the base of a settling tower. Retention of turgor is insured, by inserting leaf petioles in small vials of water. Following treatment, a plant or leaf is placed in a cage where it is exposed to insect attack. The effectiveness of the repellent is determined by the number of eggs deposited or the extent of leaf area consumed. Exact measurements of leaf area may be procured directly by means of a planimeter, by photoelectric determination, or by estimation with the aid of cross-section paper.

On a par with the volume of effort directed toward the study of repellents for use against plant-feeding insects is that of the evaluation of mosquito repellents. The accepted system consists of applying 1 cc. of repellent to the forearm from wrist to elbow and introducing the arm into a cage containing from several hundred to several thousand mosquitoes. Repellency is determined by the time to the first bite measured in minutes and by the biting rate per 30 seconds. Many variables enter into this test and must be controlled in order to yield reproducible results. The outstanding causes of variation are the differences in different batches of a given chemical due to impurities, oxidation and other changes in test chemicals, the effects of different diluents, possible knockdown effects on mosquitoes, physiological factors in mosquitoes, the species used, environmental conditions such as temperature and humidity, the different attractive value of each subject, and the amount of sweating. The success of evaluation depends upon standardization of variables of this sort. Those originating in the chemical are the most easily controlled. The effects or synergistic action of diluents can be predetermined and controlled. Probably greater attention should be paid to the physiological age of the insects. There are objections to testing various repellents on the same group of

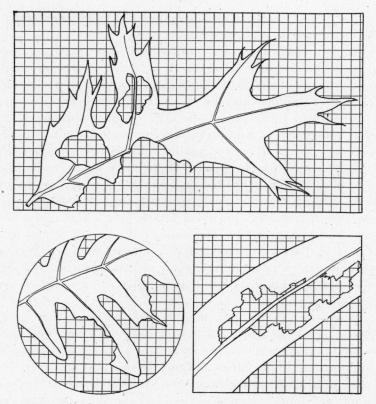

Fig. 61. Method of estimating leaf area consumed by insects. (Modified after Hansberry, "Laboratory Procedures in Studies of the Chemical Control of Insects," *Am. Assoc. Adv. Sci.*, 1943.)

mosquitoes as some repellents have knockdown or anesthetic properties. In addition, the element of fatigue is introduced. The most difficult variables to control are those arising from the unequal attractive values of different subjects and the same subject at different times. Pairing a candidate repellent with a standard such as dimethyl phthalate tends to eliminate subject variation but possesses other drawbacks. One such drawback is the possibility of one compound exerting on the insect an effect which might modify its response to the other member of the pair.

Repellents Extracted from Plants. In his search for chemical repellents man turned first to the plant kingdom. Here he had discovered fish and arrow poisons. Subsequently, several of these were adapted for use as insecticides. Of those that possess repellent as well as insecticidal properties four of the most famous recovered from this source are pyrethrum, derris, nicotine, and oil of citronella. Pyrethrum is the name applied to dried flowers or extracts of dried flowers of *Chrysanthemum cinerariaefolium* Trev. Its toxic principles are pyrethrins 1 and 2. As pyrethrum is an effective insecticide not much emphasis has been placed upon its repellent qualities although these are of high order. Against full-grown codling moth larvae the substance is effective when made up as a 5 per cent extract plus 5 to 10 per cent cottonseed oil emulsified with blood albumin (Yothers and Carlson, 1944). It is soaked in bands of corrugated paper which when dry are placed around tree trunks. Holden and Findlay (1944) found that pyrethrum in a vanishing cream base gave six-hour protection from *Glossina palpalis* in the Gold Coast, provided sweating was not extreme. It was effective also against *Culicoides*. Although it did not prevent flies from settling on the skin, it did prevent them from biting, thus indicating its action as a gustatory repellent. Similar results were obtained by Hornby and French (1943). Pyrethrum-thiocyanate oil sprays have been used on cattle with some success. Such sprays, according to Howell and Fenton (1944), give four-hour protection against *Stomoxys calcitrans* (L.) and $10\frac{1}{2}$ hours against *Haematobia irritans* (L.).

Derris is a generic term for the roots of species of *Derris*, *Lonchocarpus*, *Tephrosia*, etc., in which the chief toxic principle is rotenone ($C_{23}H_{22}O_6$). Nicotine, from *Nicotiana tabacum*, is β-pyridyl-α-n-methyl-pyrrolidine. From *Anabasis aphylla* L. is derived anabasine which is considered an isomer of nicotine. Oil of citronella, long famous as a mosquito repellent and used as early as 1882, is an essential oil extracted from *Andropogon nardus* L. From 55 to 92 per cent of the oil is geraniol. Citronellol, citronellal, borneol, etc., constitute the balance.

The continued search for naturally occurring plant repellents centered around those plants that were normally avoided by polyphagous insects. More often than not, however, it was discovered that many substances which because of their repellency conferred immunity upon a plant were ineffective when they had to compete against the strong attractiveness of preferred plants. Investigations along these lines involved two procedures: extraction and isolation of the repellent, and testing of its effectiveness against specified insects. Generally speaking, the usual methods of extraction of plant materials were followed, namely, steam distillation, maceration and extraction with solvents, and enfleurage. Extracts to be tested were placed in olfactometers to determine absolute repellency, or were sprayed upon an attractive food or oviposition site. In the latter, repellency was measured by the amount in area of food consumed (Fig. 61) or the amount, in number of eggs or larvae deposited, of oviposition.

In an attempt to extract effective repellents from plants immune to attack by the Japanese beetle Metzger and Grant (1932) extracted 390 species (326 genera and 108 families). Of these, only 56 showed any indication of repellent qualities. In the case of dried material extraction was accomplished by maceration and percolation in an alcohol-water solvent. Fresh material, after being finely chopped, was allowed to stand overnight, after which period it was centrifuged. For testing, extracts were sprayed on representative orchard trees and the number of beetles per tree counted as compared to those found on check trees sprayed with water only. This is the usual procedure followed in testing repellents in the field.

Other naturally occurring repellents include extracts of the tropical plants, *Tephrosia* and *Lonchocarpus*, which are extremely repellent to young larvae (Gimingham and Tattersfield, 1928), extracts of leaves of the tree *Melia azedarach* L. which protect garden crops against ravages of swarming North African grasshoppers, an essential oil in the tropical grass *Melinus minutiflora* which by its odor repels mosquitoes and ticks, thus protecting cattle which invade pastures of this grass (Morgan, 1940), and steam distillates of *Tagetes minima* which is repellent to blowflies (Mönnig, 1936).

Synthetic Repellents. With the accidental discovery that certain commonly employed refined and synthetic materials repelled insects, serious attempts were directed toward isolating powerful repellent compounds. It has been largely a hit or miss proposition. The outstanding products that have been developed are listed in the company of successful repellents of plant origin in the paragraphs that follow.

Plant-feeding Insects. To the Japanese beetle the following sprays have been found to be repellent: three applications of 20 lbs. of hydrated lime plus 3 lbs. of aluminum sulfate to 100 gals. of water sprayed on apples (Metzger and Lipp, 1937); derris, the repellent action being attributed to rotenone and deguelin (Fleming and Baker, 1936); phenolthiazine on grape foliage (Johnson, 1941); tetramethylthiuram disulfide (Pierpont, 1939; Guy and Dietz, 1939). The last mentioned is effective also against the Colorado potato beetle, the Mexican bean beetle, and the eastern tent caterpillar (Guy, 1936). Derris repels the Mexican bean beetle, *Epilachna varivestis*. Bordeaux mixture and derris powder is effective against many insects (Huckett, 1941). Whitewash, a mixture of $Ca(OH)_2$ plus $ZnSO_4$ as a binder for the lime plus blood albumin or casein (a spreader), when sprayed on entire trees repels over-wintering adults of the potato leaf hopper, *Empoasca fabae* Harris (McDaniel, 1937; Woglum and Lewis, 1940; Lewis, 1942). Bordeaux mixture plus nicotine sulfate repels the melonfly, *Dacus cucurbitae* (Marlowe, 1940). An emulsion of pyridine, naphthalene, or creosote sprayed on beets and cauliflower will repel and prevent oviposition by pests of these crops (Mesnil, 1934). To the Queensland fruit fly, *Chaetodacus tryoni*, sodium fluosilicate is only slightly repellent but molasses definitely so (Allman, 1940). Oviposition by oriental fruit moths can be reduced 50 per cent by the use of furfural, Dippel's oil (from the destructive distillation of bone—most important constituent is pyridine), ammonium sulfide, or amyl acetate (Lipp, 1929). Flea beetles injure seedling turnip leaves only; by treating seeds with paraffin or turpentine, Jenkins (1928) was able to reduce injury to 3.2 per cent. Apparently the repellent odor is transmitted to the seedling leaves. Brinley (1926) found that copper stearate and copper resinate, although comparatively nontoxic, seemed to repel tent caterpillars in that feeding was inhibited and starvation resulted.

Honey Bees. One of the recognized evils of widespread spraying is the killing of many beneficial insects along with the pests against which the spray is specifically directed. Chief among the victims is the honey bee. To prevent bees from imbibing poisoned spray, repellents have been tested which could be mixed with the spray. Repellents likewise are being sought that will prevent bees from pollinating because it is sometimes desired to prohibit heavy stands of fruit. Thus far no effective repellents have been developed that do not slightly injure the blossoms. Shaw (1941) tested the action of creosote, tar oil, phenol compounds, naphthalene, and alpha and beta naphthols. These reduced the number of visits by bees but they also slightly injured the blossoms. According to Turnbull (1945), however, a solution of 2 to 4 oz. of carbolic acid in 100 gal. of water is quite satisfactory.

Mosquitoes and Blood-sucking Flies. The greatest field for repellents lies in the prevention of attacks on man by blood-sucking insects. No greater need exists than that of preventing vectors of diseases such as malaria, trypanosomiasis, filariasis, pappataci fever, and others from biting. As long as man must live in areas where these diseases are prevalent there is urgent need for a good repellent. The theoretical criteria for an *ideal* repellent, as stated some years ago, are: (1) complete protection of the treated area for several hours under all conditions of weather and infestation, (2) protection against all biting insects, (3) lack of toxicity and irritation to skin and mucous membranes, (4) lack of unpleasant odor, (5) harmlessness to clothing, (6) possession of an esthetic touch and appearance, (7) ease of application, and (8) cheapness.

Experience during World War II showed that the most important criterion is cosmetic acceptability. Regardless of how perfectly a repellent meets the criteria enumerated above it is worthless if it is not used. Troops in tropical theaters of war applied repellents only under duress because the formulations currently issued burned areas of the face, dissolved plastics so that watch crystals, cigarette cases, and flashlight holders became too sticky to handle, and were unpleasant under conditions of extreme heat, humidity, and sweating.

Second in importance to cosmetic acceptability is repellency. A substance must not only possess intrinsic repellency; it must be capable of offsetting the attractive stimuli of man. At the moment no standard method of testing the intrinsic repellency of substances against mosquitoes has been put to widespread use. It is important that a method be devised by means of which the repellency of candidate materials can be tested first against an artificial standard attractant. A test of this sort would reveal whether the candidate acted as a gustatory or olfactory repellent. Some substances function in one manner for one species and in a different manner for others. Some insects are more effectively repelled by a gustatory than an olfactory repellent and vice versa. Secondly, this test would indicate the time of repellency uncomplicated by absorption and other factors. It might indicate a consistent correlation between boiling point (or some other constant) and effectiveness or duration of repellency. Up to the present repellency has been tested on human subjects as already described. This procedure has served to screen large numbers of compounds during a period when speed was necessary. We still know little of the phenomenon of repellency.

Having determined absolute repellency against a standard, it then behooves us to test candidate compounds on human subjects to ascertain the effects on duration of repellency of the various chemical, physical, and physiological characteristics of the host. Of these, the more important are sweating, creep, absorption (rather than evaporation or oxidation) by the skin, mechanical rubbing, and natural host atttraction.

The tendency of sweat to impair repellency has been studied extensively at the Naval Medical Research Institute, Bethesda, Maryland by Pijoan and Jachowski (1945) and Jachowski, Pijoan, Blodgett and Gerjovitch (1945). Here the testing methods employed were fundamentally similar to those described on p. 209. When it became apparent that many repellents which were efficient under these conditions failed to afford satisfactory protection in the field, the policy was adopted of subjecting repellents to an initial screening by this method (i.e., on dry skin) and selecting the promising ones for further testing under simulated tropical conditions. In the first instance a room temperature of 80° to 84° F. dry bulb and 68° to 72° F. wet bulb was maintained since it approached optimum breeding conditions and was below the sweat threshold of subjects. In the second case the room temperature was set at 80° to 90° F. dry bulb and 80° to 82° F. wet bulb. To produce a standing sweat subjects were exercised mildly. As an added precaution they were kept in the hot room for 30 minutes prior to the application of repellent for the purpose of allowing their arms to come to equilibrium with the environment. Tests were run with *Aedes aegypti* under a normal biting rate of nine bites per 30 seconds. Rutgers #612 was tested daily as a control.

These tests yielded readily reproducible results and data which checked remarkably closely with those acquired by tropical field testing. In addition, they emphasized the importance of sweat as a factor decreasing repellency time, and revealed the failure of heretofore acceptable repellents to operate efficiently under tropical conditions. Dimethyl phthalate and Rutgers #612, for example, failed to repel for more than two hours. Other, more acceptable, compounds were developed.

Creep, another important factor in the duration of repellency, may be defined as a progressive wetting expedited along the sharp reëntrant angles of the skin sulci by capillarity (Christophers, 1945). Initially, gravity is of little import. Viscosity is the dominant factor in determining the value of creep index, hence the extent of creep varies appreciably with different materials. Creep index may be defined as the factor by which in a given period of time the area of spread exceeds an original area of two square inches treated with a given amount of material.

Early repellents were chiefly essential oils (e.g., citronella, eucalyptus, pennyroyal, rose geranium, cedarwood, thyme, wintergreen, clove, lavender, cassia, anise, bergamot, pine tar, bay laurel, sassafras), certain acetates (e.g., linalyl, santalyl, terpinyl, geranyl), glycols, indalone, and pyrethrum. Dimethyl phthalate proved to be the most successful repellent and has been used as a standard against which to test candidate repellents. The best all-round repellent against mosquitoes and biting flies in use during World War II was a mixture of dimethyl phthalate, 2-ethyl-hexane-diol 1, 3 (Rutgers #612), and indalone (alpha alpha dimethyl alpha carbobutoxydihydro gamma pyrone) in a proportion of 6:2:2. Thousands of substances have been tested since. To date no consistent correlation has been found between repellency and chemical formula or configuration. Two of the most promising groups are certain hydrogenated naphthol derivatives and hydrogenated diphenyls (Pijoan, *et al.*, 1945). The mixing of these with 2-phenyl cyclohexanol results in a prolongation of repellency as a result of synergistic interaction. Three of the most effective, which repel *Aedes aegypti* for longer than 289 minutes at environmental temperatures of 90° F. dry bulb, 80° F. wet bulb, are: 2-phenyl cyclohexanol (washed) plus 2-naphthol, 1,2,3,4 tetrahydro, acetylglycine ester; 2-phenyl cyclohexanol (washed) plus 2-naphthol, 1,2,3,4 tetrahydro, glycollic ether (crude); and 2-phenyl cyclohexanol (washed) plus 2-cyclohexyl cyclohexanol (washed). Since, however, toxicological studies remain to be done on a number of these compounds their use as repellents cannot be recommended as yet.

Synergistic and Antagonistic Action of Solvents on Repellents.
During the course of biological tests of solutions and mixtures com-

posed by adding repellents and solvents to a cosmetically satisfactory plastic base in such a manner as not to alter the physical
properties of the film it was noted that there was a decreased action
of the repellent not explainable on the basis of dilution (Pijoan
and Jachowski, 1945; Jachowski, et al., 1945). Further investigation
of the interaction of repellents and other ingredients revealed,
moreover, that often there existed a reverse phenomenon. For
example, *Aedes aegypti* was repelled 10 times longer by dimethyl
phthalate and hexyl alcohol than by dimethyl phthalate to which
had been added ethyl alcohol as a solvent. This led to the deduction
that an insect repellent when in solution or mixture may be modified by (a) a quantitative reduction of repellent by dilution, (b) the
repellent properties of the solvent, and (c) the repellent effect
induced by the new physical and chemical properties of the solution.

For the purpose of evaluating such solutions (some were mixtures,
but for convenience ingredients are termed solute and solvent
and their combination termed mixture) with reference to an
increase in the action of the repellent (synergism) by the solvent
or an inhibition of repellent activity (antagonism) the Synergic
Index was established (Pijoan and Jachowski, 1945). This Index
compares the actual repellent time of a mixture to the theoretical
duration calculated from the effectiveness of each ingredient. The
Index assumes that the final preparation is a mixture.

Since the function of a mixture is to prolong the length of time
the mosquitoes are repelled, the values T solute, T solvent, and T
mixture, which represent respectively the time interval from application to first mosquito bite of pure solute (repellent), pure solvent,
and of the resulting mixture of these in any given proportions, are
important. Furthermore, it is necessary to know the fraction (f) by
volume or by weight/volume of the solute in a mixture.

When the repellent mixture is more effective than the repellent
alone as far as the protection period is concerned, the increased
effect is a simple ratio:

$$\frac{T \text{ mixture}}{T \text{ solute}} = \text{Repellent Index (RI)}$$

It is important to note, however, that the effect of proportion or
dilution to the solvent is not considered. Neither can the theoretical
value be determined since the phenomenon of interaction is not
usually additive. Thus T mixture is not ordinarily equal to T solute

$+ (1 - f)$ T solvent but can be related to an interaction in terms of actual proportional repellent times of its components, i.e.,

$$\frac{\text{T mixture}}{\text{T solute} + (1 - f) \text{ T solvent}} = \text{Synergic Index (SI)}$$

An SI greater than 1 indicates that the interaction is in the direction of increased repelling time. If SI is less than 1, a converse interaction is indicated. These values bear significant relation to the proportion of solute and solvent. In either case it should be noted that the value of SI is a measure of degree of interaction rather than of effectiveness of solution. SI denotes degrees of synergism or antagonism of mixture ingredients in either the solvent or the solute, and solutions once made can be rated for practical use by RI which in turn can be calculated from SI. The relation existing between the two measures, SI and RI, is shown in the following equation:

$$SI = \frac{\dfrac{\text{T mixture}}{\text{T solute}}}{f + (1 - f) \dfrac{\text{T solvent}}{\text{T solute}}} = \frac{(RI)}{f - (1 - f) \dfrac{\text{T solvent}}{\text{T solute}}}$$

Thus:

$$RI = (SI)f + (1 - f) \frac{\text{T solvent}}{\text{T solute}}$$

In choosing solvents for use in repellent mixtures the SI is of indispensable aid in deciding upon the type of solvent and its proportion in the mixture.

Impregnated Clothing and Fabrics. Because fabrics retain repellents for much longer periods of time than does human skin, the impregnating of clothing with repellents has proved very successful as a means of protection against blood-sucking insects. Twill or open-mesh shirts sprayed with 50 to 100 cc. of liquid have afforded complete protection against mosquitoes for one week and partial protection during the second week under extreme test conditions of 7200 bites per minute. Nets of a mesh ordinarily large enough to admit mosquitoes also will afford protection if impregnated with dimethyl phthalate. Fish nets of 1.1 cm. mesh and 0.5 mm. strand diameter impregnated with 0.32 cc. of dimethyl phthalate per gram of net have been shown to combine the most desirable features for protection and comfort in the tropics. The success of simple impregnation of this sort led to extensive experiments with solid as well as liquid repellents, but details have not yet been released. The incorporation of vinylite binders has lengthened the duration of repellency, especially under adverse conditions of wear and laundering.

The principle of substituting wide mesh for small mesh nets by impregnation has been applied with equal success to house screens. Hull and Shields (1939) recommended brushing screens with either (1) one part concentrated pyrethrum (extract of 20 lbs. of pyrethrum in 20 gals. of refined kerosene) plus 20 parts of lubricating oil, or (2) one part of concentrated pyrethrum plus 6 parts of kerosene plus 12 parts of lubricating oil to prevent the entry of *Culicoides*.

Ticks and Chiggers. Although adult ticks generally exhibit no reaction in the presence of repellents, these insects in the immature stages can be effectively repelled. Dimethyl phthalate and indalone will reduce attachment by 50 to 60 per cent in 24 hours. Rotenone and sulfur are equally effective but cause mild dermatitis in some people. It has been found that treating the margins of clothing is just as effective as treating entire surfaces. In Russia, mixtures of turpentine, mint leaf infusions (*Mentha piperita* and *M. crispa*), wood tar, ortho- and metachlorphenol, and acetic acid are said to be effective against ticks in all stages of development (Moskvin, 1939).

The success of benzyl benzoate as a miticide has minimized the necessity of developing repellents against these arthropods. Dimethyl phthalate, also a miticide (Madden, Lindquist, and Knipling, 1944), is said to possess some repellent properties.

Repellents also have been sought for use against ticks infesting livestock. A new method for testing the efficiency of candidate tick repellents on sheep has been devised by Brody (1939). It involves the use of an adjustable head gear (see Fig. 62) and two cylindrical cages fastened in place, one over each ear. Empty cages are attached one week following treatment of the ears with candidate repellents. When the animal is considered to have become accustomed to the apparatus, usually after two days, vials containing ticks of both sexes are inserted in a vial holder on each cage. Ticks have free access to the ears from these vials. Repellency is calculated from the number of ticks dead, attached, and engorged.

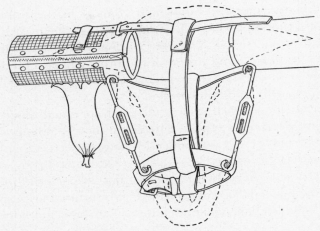

FIG. 62. Diagram of ear cage and halter used in evaluating candidate repellents against the gulf coast tick. (Redrawn after Brody, *U.S. Dept. Agr. Circ.*, 1939.)

Blowflies. The purpose of blowfly repellents is to safeguard sheep and cattle against attack (strike) by species of blowflies. The requirements differ somewhat, depending upon the type of fly, the kind of animal being protected, and the region of the world involved. To be effective a dip or spray should possess the following characteristics (Moore, 1937):

1. Repellency
2. Long persistence on wool or hair and ability to withstand rain
3. Ability to wet and penetrate fleece and hair
4. High antiseptic properties to prevent bacterial action
5. Nonstaining properties
6. Cheapness

Potential repellents may be tested in the laboratory or in the field. Parman, *et al.* tested repellency by the ability of the product to counteract the attractiveness of meat; Hobson (1937), by the ability to counteract the attractiveness of ammonium carbonate and indole in fleece.

Field testing formerly meant dividing a herd or flock into check groups and sprayed groups. Because susceptibility, or host attractiveness as it is termed with respect to human beings, is exceedingly complex and variable, this method was not valid. Factors contributing to complexity and variability are weather, hair color, breed, time of day, etc. Variation in fly count may be nonrandom or random (experimental). The former is due to breed, time of day, etc.; the latter, to animal differences. Fryer, *et al.* (1943) and Shaw, *et al.* (1943) studied these objections and attempted to establish a standard technique for screening.

Previously, the lack of standard technique existed because of (1) the many variables, (2) the type of data obtained, (3) the lack of statistical methods suitable to this type of experimentation. Two general techniques now in use for testing fly sprays under field conditions are the whole-cow method and the half-cow method. The former is the more accurate. Objections to the half-cow method include the false assumption that flies which are driven from the sprayed side of the animal go to the unsprayed side, the interference of spray odors and head and tail movements with the status of repellency, inadequate statistical methods, and lack of uniform formulae for measuring repellency.

To set up the whole-cow method groups of animals are composed by balancing susceptibility. With care in the formation of such groups the same number of flies within measurable experimental variation is attracted to each group. Each group is then rotated from spray to spray to check so that eventually all sprays have been applied to each animal. This plan may be accomplished by the application of a randomized latin square design on balanced groups of animals and different time periods (for basic latin squares see Fisher, 1937 and Fisher and Yates, 1938).

Since periodic counts of flies present on animals are distributed in a Poisson-like manner, statistical data based on normal distribution are not valid unless normalized. Various formulae have been worked out for transforming data (Fryer, *et al.*, 1943). Using this technique Shaw, *et al.* found Thanite (fenchyl thiocynyl acetate) to be a very effective spray for cows.

Generally speaking, the best repellents are (1) products of the destructive distillation of the long leaf pine, *Pinus palustris*, namely, pine tar, pine tar oil, wood naphtha whose action is probably due to phenols, hence also phenols, (2) pyrethrum, (3) strong inorganic antiseptics such as mercuric chloride and copper compounds. As powders, copper compounds are even more effective (on beef liver) than strongly odorous compounds. It is assumed that this power is due to their ability to alter the decomposition of meat so that the products evolved are unattractive. Mercuric chloride probably is a gustatory repellent in itself. Deonier (1939) has shown that adults of *Cochliomyia americana* and *Phormia regina* with their tarsal receptors can detect and are repelled by solutions of mercuric chloride.

Table 39

SOME MIXTURES RECOMMENDED FOR USE AGAINST BLOWFLIES

Formulations	Authority
Pine tar oil	Laake, *et al.*, 1926
Glycerol diborate and 25% alcohol	McGovran and Ellison, 1936
0.3%	
Oil of cloves	
Oil of pennyroyal	
Oil of wintergreen	
Eugenol	
Safrol	
Isoquinoline	
Chlorophenols	
Benzol chloride	
Benzyl chloride	
α-Chloronaphthalene	
0.9%	
Oil of citronella	
Pine tar	
Tar	
Paraffin	
Phenol	
Cresols	
Methyl salicylate	
Dimethylaniline	
p-Xylidine	
α-Naphthylamine	
Crotonaldehyde	
Pinene	Hobson, 1937
Mixture of:	
Water-white mineral oil (38 pts. by wt.)	
Pale straw colored cresylic acid (25)	
p-Dichlorobenzene (25)	
Crude brown Yorkshire wool grease (12)	
4 pts. by vol. of oleum pices rectified to every 66 parts of the above	Moore, 1937
Olive oil, cottonseed oil, oleic acid	Hobson, 1940
3% phenoxychloroethyl ether in highly refined oil	Searls and Daehnert, 1941
Methallyl disulfides in a carrier of carbitol acetate and polymerized butylene	Loeffler and Hoskins, 1943
Thanite (fenchyl thiocynyl acetate)	Shaw, *et al.*, 1943

Houseflies. Kilgore (1939) tested the comparative repellency of materials to *Musca domestica* by the sandwich-bait method. Uniform baits were prepared with brown molasses applied to strips of white blotting paper over which was placed a porous paper with the chemical to be tested. Best of those tested were diethylene glycol monobutylether acetate, amyl salicylate, triethanolamine, and butylmesityloxide. Ralston and Barrett (1941) found the following to be effective: decyl, undecyl, undecenyl, and dodecyl alcohols and aliphatic nitriles (with 10 to 14 carbon atoms) of fatty acid derivatives. Undecylonitrile, lauronitrile, tridecylonitrile, and dodecyl alcohol were more repellent than oil of citronella. Moore (1934), after testing several saturated and unsaturated hydrocarbons, esters, and alcohols, found that repellency increased from hydrocarbon to alcohol to ester. The slightly volatile unsaturated cyclic ester, santysyl acetate, was best.

Termites and Other Wood-boring Insects. The food of termites is principally cellulose which has been invaded by fungi. Wood-dwelling members of the order attack wood directly from the outside, but earth-dwelling species enter from the ground or from closely adjacent pieces of wood. Many show very marked food preferences Numerous striking examples are cited by Kofoid and Bowe (1934). These authors and Williams (1934) found that selections are determined by different factors in the case of each species of termites. Outstanding factors are: moisture content (both free and bound water) of the wood, amount and chemical nature of extractives, physical qualities affecting the density, the hardness, and the rate of growth, differences between heartwood and sapwood, the nature and amounts of nutrient substances which favor the growth of fungi, the nature and extent of pre-existing fungus attack, and the contents of wood affecting taste and food value. The nature of repellence of wood or, more broadly speaking, its termite resistivity is a combination of all these factors plus others still unknown.

Since one approach to the control of termites involves the use of woods that are naturally resistant to termite attack, it is necessary to test termite resistivity. Candidate repellents and preservatives are tested in the same manner. In general, there are two procedures: one, the so-called graveyard method, is based on preferences; the other makes use of nonpreferential controlled tests. In the graveyard test, stakes are placed in ground heavily populated with termites and observations made as to which is most heavily attacked. As food preference is also a function of availability field tests of this sort are largely empirical.

Data of a more absolute nature are yielded by nonpreferential tests. Here termites are sealed in a hollowed part of a test block. After a given time lapse under carefully controlled conditions the extent of excavation is determined by placing the block in a vacuum, filling the cavities with mercury, and measuring the amount of mercury required. With this method Kofoid and Bowe found that there was a direct correlation between the amount of extract and the volume of wood excavated.

Extractives act by repelling, by killing, or by inhibiting the growth of wood-destroying fungi. Some woods, like Port Orford cedar, have much of the extractive in the form of a volatile oil which affords the principal source of resistance. Termites will attack wood from which this oil has been removed by weathering in nature or continuous heating in the laboratory. High resin content also seems to influence palatability adversely. In view of this it is not surprising that compounds extracted from conifers and from the destructive distillation of coniferous wood are especially successful as repellents. Wood tar and coal tar crudes are used widely. In general, covering or impregnating wood with liquid coal tar creosote affords some protection. Wolcott (1945) found that the dry-wood termite, *Cryptotermes brevis* (Walker), is repelled by heavily chlorinated or brominated phenols specifically hexachlorophenol, pentachlorophenol, copper pentachlorophenate (the most effective), fluorene, phenanthrene, fluoranthrene, and pyrene.

Second only to termites as destroyers of wood are bark beetles. Because of the importance of many species as vectors of fungus diseases which kill trees or render lumber unsaleable concentrated effort has been directed toward beetle control. Among the more useful toxic and repellent sprays developed are those containing naphthalene, orthodichlorobenzene, monochloronaphthalene, diphenyl flakes, and pentachlorophenol. Sprayed on logs these afford protection against *Scolytus multistriatus* and *Hylurgopinus rufipes* (Whitten, 1942).

Subterranean and Root-dwelling Insects. Crystalline coal tar hydrocarbons have been used very successfully against insects in this category. The great repellent power of some of these compounds may be due to the fact that a few such as naphthalene and anthracene are not found in nature. Naphthalene flakes applied three or four times a season at the rate of 1 lb. per 100 ft. of row is recommended as a carrot rust fly repellent (Whitcomb, 1938). Larvae of the narcissus fly, *Meroden*, are repelled from greenhouse beds of coke and crude naphthalene. Sulfur seems to prevent oviposition in the sod by June beetles, *Phyllophaga anxia* Lec. (Hammond, 1929). Lipp (1934) found that paradichlorobenzene prevented oviposition by the Japanese beetle in an experimental pot for 32 days. Blunk (1938), however, maintains that there are no satisfactory means of controlling larvae of May beetles without also harming the land.

Mothproofing. Fabrics may be mothproofed by treatment with volatile repellent materials or by impregnating the fiber with a nonvolatile substance. The former method is obviously the less satisfactory of the two. Ninety per cent of clothes moth repellents of a nonvolatile nature are either sodium silicofluoride or combinations of sodium silicofluoride and Al, Mg, or NH_4 (Van Antwerpen, 1941). The process is similar to dyeing. Certain urea derivatives also act successfully, namely, allylthiourea, thiourea, phenylthiourea, and *o*-tolylthiourea. Another effective compound is dixylylguanidine dissolved in naphtha. When the naphtha evaporates, a repellent amorphous mass remains.

Miscellaneous. Packaged goods frequently are subjected to severe attack by household or factory pests chief among which are roaches, silverfish, and firebrats. Sweetman and Bourne (1944) developed a two-ply laminated Kraft paper sealed with an asphalt-glue adhesive which offers resistance to penetration and feeding by the American, Australian, German, and brown-banded roach, the silverfish, the four-lined silverfish, and the firebrat. A different technique directed to the same end was developed as a direct outcome of fundamental work on taste threshold by Frings (unpublished). The knowledge that certain salts were refused at very low concentrations was utilized to impregnate packaging material. The nontoxic nonodorous properties of these salts are distinct advantages. The disadvantage of their ready solubility in water may be overcome by impregnating the packaging material with wax.

Hazard (1945) reported that a surfacing cement of finely divided copper powder incorporated into magnesium oxychloride cement is repellent to roaches. The mechanism of action is not known.

Discussion. Our knowledge of the mode of action of chemical repellents may be summarized by the single neutral statement that some compounds repel and others do not. All efforts to relate repellency to chemical structure have been fruitless. Neither purposeful study nor the perusal of lists of thousands of repellents has offered productive clues. There appears to be little selectivity in the phenomenon of repellency. While some evidence has been accumulated to indicate that there may be a relation between boiling point and degree of repellency and between the presence of oxygenated radicals and repellency as well, the fact of the matter is that these relationships exist with *response* to chemicals irrespective of whether it is positive or negative. Boiling point is a measure of volatility. The first prerequisite of an olfactory repelleint is that it must be volatile. The second is that the molecules must be able to penetrate the cuticle and cell membranes of an end organ after which they must be adequate stimuli. Given these conditions the final determining factor is a genetic one in that the genetic constitution of an animal determines whether the final response shall be positive or negative.

In short, an understanding of the phenomenon of repellency is more certain to be forthcoming if greater attention is paid to the fundamental aspects of chemoreception. Voluminous probing in the dark has, it is true, produced economically effective formulations, but basic knowledge of repellency has not been advanced with the result that the direction of future developments cannot be planned.

REFERENCES

Allman, S. L.: Foliage poisons for the Queensland fruit fly (*Strumeta tryoni* Froggatt). The repellent effect of molasses, *J. Australian Inst. Agr. Sci.*, **6**(3), 154–160 (1940).

Anonymous: Test insecticide repellency, *Soap Sanit. Chemicals*, **16**(7), 127 (1940).

Anonymous: Quarterly Report of the Consultant Malariologist, ALFSEA, 1945.

Blagoveshchensky, D. I., N. G. Bregetova, and A. S. Monchadsky: New deterrent substances for protecting man against attacks of mosquitoes. *Compt. rend. acad. sci. U.R.S.S.*, N.S., **40**(3), 119–122 (1943).

Blunk, H.: Das Schriftum über die Möglichkeiten zur Bekämpfung der Maikäferengerlinge mit mechanischen und chemischen Mitteln, *Z. Pflanzenkrankh. Pflanzenschutz*, **48**(2), 64–87 (1938).

Brinley, F. J.: Insecticidal value of certain war chemicals as tested on the tent caterpillar, *J. Agr. Research*, **33**(2), 177–182 (1926).

Brody, A. L.: A method for testing the value of chemical mixtures as repellents of the Gulf Coast tick, *U.S. Dep. Agr., Bur. Entomol. Plant Quarantine, Circ.* ET-152, 1–14 (1939).

Bunker, C. W. O., and A. D. Hirschfelder: Mosquito repellents, *Am. J. Trop. Med.*, **5**(5), 359–383 (1925).

Buxton, P. A.: Ministry of Production: Pyrethrum Development Panel, British Work on the Control of Biting Insects, 1943.

Cameron, T. W. M.: Fly sprays and repellents, *Can. J. Comp. Med.*, **5**(10), 275–282 (1941).

Christophers, R.: Ministry of Production: Insecticide Development Panel, Mosquito Repellency Inquiry, 1945.

——: Insect repellents, *Brit. Med. J.*, **3**, 222–224 (1945a).

Clark, C. O.: The protection of animal fibers against clothes moths and dermestid beetles, *J. Textile Inst.*, **19**(12), 295–320 (1928).

——: A history of commercial mothproofing, 1920–1940; *J. Soc. Dyers Colourists*, **59**(10), 213–215 (1943).

Denham, C. S.: The acid secretion of *Notodonta concinna*, *Insect Life*, **1**, 143 (1888).

Deonier, C. C.: Responses of the blowflies, *Cochliomya americana* C. and P. and *Phormia regina* Meigen, to stimulations of the tarsal receptors, *Ann. Entomol. Soc. Am.*, **32**(3), 526–532 (1939).

Downes, W.: Recent trials of repellents for Narcissus fly, *Can. Entomol.*, **67**(2), 21–24 (1935).

Fisher, R. A.: "Design of Experiments," 2d ed., London, Oliver & Boyd, 1937.

——, and F. Yates: "Statistical Tables for Biological, Agricultural, and Medical Research," London, Oliver & Boyd, 1938.

Flanders, S. E.: Olfactory responses of parasitic Hymenoptera in relation to their mass production, *J. Econ. Entomol.*, **37**(5), 711–712 (1944).

Fleming, W. E., and F. E. Baker: Derris as a Japanese beetle repellent and insecticide, *J. Agr. Research*, **53**(3), 197–207 (1936).

Fredericq, L.: Die Sekretion von Schutz- und Nutzstoffen. VIII. Arthropoda, *Handb. vergl. Physiol.*, **2**(2), 112–166 (1910).

Freeborn, S. B.: Observations on the control of Sierran Aëdes (Culicidae: Diptera), *Pan-Pacific Entomol.*, **4**(4), 177–181 (1928).

Frings, H.: (Personal communication, 1946.)

Fryer, H. C., A. O. Shaw, F. W. Atkeson, R. C. Smith and A. R. Borgmann: Techniques for conducting fly-repellency tests on cattle, *J. Econ. Entomol.*, **36**(1), 33–44 (1943).

Fulton, B. B.: Physiological variation in the snowy tree-cricket, *Oecanthus niveus* De Geer, *Ann. Entomol. Soc. Am.*, **18**(3), 363–383 (1925).

Garb, G.: The eversible glands of a chrysomelid larva, *Melasoma lapponica*, *J. Entomol. Zool.*, **7**(1), 88–97 (1915).

Gerjovich, H. J., and M. L. Hopwood: Directions for purification of commercial grade compounds for the insect repellent NMRI-201, Research Project X-168, Report No. 7, Naval Med. Research Inst., Aug. 25, 1945.

Gimingham, C. T., and F. Tattersfield: Laboratory experiments with non-arsenical insecticides for biting insects, *Ann. Applied Biol.*, **15**(4), 649–658 (1928).

*Goetsch, W., R. Grüger, T. Latosek, and K. Offhaus: Control of ant colonies, *Z. angew. Entomol.*, **29**, 219–242 (1942).

Granett, P. J.: II. Relative performance of certain chemicals and commercially available mixtures as mosquito repellents, *J. Econ. Entomol.*, **33**(3), 566–572 (1940).

Guy, H. G.: Thiuram sulfides as repellents to leaf-feeding insects, *Ibid.*, **29**(2), 467 (1936).

——, and H. F. Dietz: Further investigations with Japanese beetle repellents, *Ibid.*, **32**(2), 248–252 (1939).

Hammond, G. H.: Sulphur as a deterrent to June beetle (*Phyllophaga anxia* Lec.) oviposition in timothy sod, *Ann. Rep. Quebec Soc. Protect. Plants*, **21,** 37–38 (1929).

Hartley, R. S., F. F. Elsworth, and J. Barritt: Mothproofing of wool, *J. Soc. Dyers Colourists*, **59**(12), 266–271 (1943).

Hazard, F. O.: Roach repellent cement, *Soap Sanit. Chemicals*, **21**(4), 129, 131, 133, 135, 157 (1945).

*Hindmarsh, W. L., and H. G. Belschner: Cutaneous myiasis (blowfly strike) of sheep. I. Glycerol diborate as a preventative of blowfly strike of sheep, *New S. Wales Dep. Agr.*, *Vet. Research Dep.*, **7,** 41–43 (1937).

Hobson, R. P.: Sheep blowfly investigations. V. Chemotropic tests carried out in 1936, *Ann. Applied Biol.*, **24**(3), 627–631 (1937).

——: Sheep blowfly investigations. VIII. Observations on larvicides and repellents for protecting sheep from attack, *Ibid.*, **27**(4), 527–532 (1940).

——: II. Recent work on the sheep maggot problem, *Ibid.*, **28**(3), 297–299 (1941).

Holden, J. R., and G. M. Findlay: Pyrethrum as a tsetse fly repellent: human experiments, *Trans. Roy. Soc. Trop. Med. Hyg.*, **38**(3), 199–204 (1944).

Hollande, A. C.: Sur la fonction d'excretion chez les insectes salicicoles, *Ann. Univ. Grenoble*, **21**(2), 459–517 (1909).

Holman, H. J.: "A Survey of Insecticide Materials of Vegetable Origin," London, Imperial Institute, 1940.

Hornby, H. E., and M. H. French: Introduction to the study of the tsetse-fly repellents in the field of veterinary science, *Trans. Roy. Soc. Trop. Med. Hyg.*, **37**(1), 41–54 (1943).

Howell, D. E., and F. A. Fenton: The repellency of a pyrethrum-thiocyanate oil spray to flies attacking cattle, *J. Econ. Entomol.*, **37**(5), 677–680 (1944).

Huckett, H. C.: Derris and the control of the Mexican bean beetle, *Ibid.*, **34**(4), 566–571 (1941).

Hull, J. B., and S. E. Shields: Pyrethrum and oils for protection against salt-marsh sand flies (Culicoides), *Ibid.*, **32**(1), 93–94 (1939).

Imperial Institute, Consultative Committee on Insecticide Materials of Vegetable Origin. 1938–1944. Quarterly Bibliography on Insecticide Materials of Vegetable Origin. Nos. 1–25, *Bull. Imp. Inst.*, 36–42..

Jachowski, L. A., and M. Pijoan: A note dealing with the effect of certain simulated tropical conditions on the activity of two mosquito repellents. Research Project X-168, Report No. 6, Naval Medical Research Institute, Aug. 1, 1945.

——, ——, W. E. Blodgett, and H. J. Gerjovich: Summary of investigations on mosquito repellents at the Naval Medical Research Institute. Research Project X-168, Report No. 3, Nav. Med. Research Inst., May 12, 1945.

James, H. C.: Repellent banding to control ants attending the common coffee mealy-bug. *Colony Protectorate Kenya Dep. Agr.*, pp. 1–14 (1930).

Jenkins, J. R. W.: Seed treatment as a means of preventing turnip flea beetle attack, *Welsh J. Agr.*, **4**, 334–342 (1928).

Jewett, H. H.: The resistance of certain red clovers and alfalfas to leafhopper injury, *Kentucky Agr. Exp. Sta. Bull.* 329, 157–172 (1932).

Johnson, H. W., and E. A. Hollowell: Pubescent and glabrous characters of soybeans as related to resistance to injury by the potato leafhopper, *J. Agr. Research*, **51**(4), 371–381 (1935).

Johnson, J. P.: Seasonal development of the Japanese beetle and spraying for the adult insect, *Connecticut Agr. Exp. Sta. Bull.* 445, 363–367 (1941).

Kemper, H.: Beiträge zur Kenntnis des Stinkapparates von *Cimex lectularius* L., *Z. Morph. Oekol. Tiere*, **15**(3), 524–546 (1929).

Kilgore, L. B.: A study of comparative repellency by the Sandwich-Bait method using confined houseflies, *Soap Sanit. Chemicals*, **15**(6), 103–123 (1939).

Kofoid, C. A.: Seasonal changes in wood in relation to susceptibility to attack by fungi and termites, "Termites and Termite Control," *University Calif. Press.*, 1934, pp. 524–531.

——, and E. E. Bowe: A standard biological method of testing the termite resistivity of cellulose-containing materials, *Ibid.*, 1934, pp. 487–513.

Laake, E. W., D. E. Parman, F. C. Bishopp, and R. C. Roark: Field tests with repellents for the screw worm fly, *Cochliomyia macellaria* Fab., upon domestic animals, *J. Econ. Entomol.*, **19**(3): 526–539 (1926).

Larson, A. O.: The host-selection principle as applied to *Bruchus quadrimaculatus* Fab., *Ann. Entomol. Soc. Am.*, **20**(1), 37–78 (1927).

Latter, O. H.: The secretion of potassium hydroxide by *Dicranura vinula* (imago), and the emergence of the imago from the cocoon, *Trans. Entomol. Soc. London*, **1892**, 287–292.

——: Further notes on the secretion of potassium hydroxide by *Dicranura vinula* (imago), and similar phenomena in other Lepidoptera, *Ibid.*, pp. 399–412.

Lea, A., and M. C. A. Nolte: Laboratory experiments on poison baits for the brown and the red locust, 1937–1938, *Union S. Africa Dep. Agr., Forestry. Sci. Bull.* 230, 55 pp. (1941).

Lesser, M. A.: Insect repellents, *Drug Cosmetic Ind.*, **48**(2), 149–153, 165 (1941).

Lewis, H. C.: Lime zinc spray as a repellent for leafhoppers on citrus, *J. Econ. Entomol.*, **35**(3), 362–364 (1942).

Lipp, J. W.: Preliminary tests with possible repellents of the oriental peach moth, *Ibid.*, **22**(1), 116–126 (1929).

——: The effectiveness of paradichlorobenzene and naphthalene in preventing oviposition by the Japanese beetle, *Ibid.*, **27**(2), 500–502 (1934).

Loeffler, E. S., and W. M. Hoskins: Evaluation of blow-fly repellents, "Laboratory Procedures in Studies of the Chemical Control of Insects," Washington, D.C., Am. Assoc. Adv. Sci., 1943, p. 173.

McCulloch, R. N.: Control of blowfly strike. Warm mixtures and jetting injury. *Agr. Gaz. New S. Wales*, **45**(12), 673–674 (1934).

McDaniel, E. I.: White coating on foliage a repellent for potato leafhopper, *J. Econ. Entomol.*, **30** (3), 454–457 (1937).

McGovran, E. R., and L. O. Ellison: Repellency of pine tar oil to wound-infesting blowflies, *Ibid.*, **29**(5), 980–983 (1936).

MacNay, C. G.: Studies on repellents for biting flies, *Can. Entomol.*, **71**(2), 38–44 (1939).

Madden, A. H., A. W. Lindquist, and E. F. Knipling: Tests of repellents against chiggers, *J. Econ. Entomol.*, **37**(2), 283–286 (1944).

Mansbridge, G. H.: On the biology of some Ceratoplatinae and Macrocerinae (Diptera, Mycetophilidae), *Trans. Entomol. Soc. London*, **81**, 75–92 (1933).

Marchal, P.: Richet's Dictionaire de Physiologie, **9**(2), 350–354 (1910).

Marlowe, R. H.: Some deterrents as a control for the melonfly, *U.S. Dep. Agr., Bur. Entomol. Plant Quarantine Circ.* E-510 (1940).

Martin, H.: "Scientific Principles of Plant Protection with Special Reference to Chemical Control," 3d ed., London, Edward Arnold & Co., 1940.

Mesnil, L.: Nouvelle méthode de lutte contre les insectes par l'emploi de substances insectifuges, *Compt. rend. acad. agr. France*, **20**(6), 30–33 (1934).

Metzger, F. W., and G. H. Grant: Repellency to the Japanese beetle of extracts made from plants immune to attack, *U.S. Dep. Agr., Tech. Bull.* 299, 1–22 (1932).

——, and Lipp, J. W.: Value of lime and aluminum sulfate as a repellent spray for Japanese beetle, *J. Econ. Entomol.*, **29**(2), 343–347 (1937).

Military Personnel Research Committee of the Med. Research Council. Progress Reports (British).

Minaeff, M. G., and J. H. Wright: Mothproofing, *Ind. Eng. Chem.*, **21**(12), 1187–1195 (1929).

Mönnig, H. O.: A new fly repellent and a blowfly dressing. Preliminary report, *Onderstepoort J. Vet. Sci. Animal Ind.*, **7**(2), 419–430 (1936).

Moore, W.: Esters as repellents, *J. N. Y. Entomol. Soc.*, **42**(2), 185–192 (1934).

Moore: The chemical control of the sheep maggot fly (*Lucilia sericata* Meigen), *Scottish J. Agr.*, **20**, 227–240 (1937).

Morgan, E.: The tropical grass *Melinus minutiflora* as a preventive against malaria and other tropical diseases, *J. Trop. Med. Hyg.*, **43**(13), 179 (1940).

Moriyama, T.: Wood ash as a repellent against insect pests, *Formosan Agr. Rev.*, **35**, 500–504 (1939).

Moskvin, I. A.: The effect of chemical irritants on the ticks *Ornithodorus papillipes* (In Russian), *Trav. acad. milit. méd., Kiroff Armée Rouge*, **18**, 59–78 (1939).

Mulhearn, C. R.: Oils suitable as vehicles for sheep blowfly dressings. Trials at Nyngan experimental farm, *Agr. Gaz. New S. Wales*, **40**(12), 905–913 (1929).

Mumford, E. P., and D. H. Hey: The water balance of plants as a factor in their resistance to insect pests, *Nature*, **125**, 411–412 (1930).

Oshima, M.: Formosan termites and methods of preventing their damage, *Philippine J. Sci.*, **15**(4), 319–384 (1919).

Painter, R. H.: The food of insects and its relation to resistance of plants to insect attack, *Am. Naturalist*, **70**(731), 547–566 (1936).

——: The economic value and biologic significance of insect resistance in plants, *J. Econ. Entomol.*, **34**(3), 358–367 (1941).

——: Insect resistance of plants in relation to insect physiology and habits, *J. Am. Soc. Agronomy*, **35**(8), 725–732 (1943).

Parman, D. C., E. W. Laake, F. C. Bishopp, and R. C. Roark: Test of blowfly baits and repellents during 1926, *U.S. Dep. Agr., Tech. Bull.* 80, 1–14 (1928).

Pavlovskii, E. N.: "Protection from blood-sucking Diptera (Mosquitoes, Midges, Sandflies, Tabanids, etc.)," (in Russian), Moscow Acad. Sci., U.S.S.R., 1941, 67 pp.

Pearson, A. M., J. L. Wilson, and C. H. Richardson: Some methods used in testing cattle fly sprays, *J. Econ. Entomol.*, **26**(1), 269–274 (1933).

Pierpont, R. L.: Japanese beetle control tests on American elm trees in Delaware, *Ibid.*, **32**(2), 253–255 (1939).

Pijoan, M., and L. A. Jachowski: A method of evaluating synergistic or antagonistic action of solvents on mosquito repellents, Research Project X-168, Report No. 2, Naval Med. Research Inst., Jan. 12, 1945.

——, ——, and H. J. Gerjovich: A mixture of two new mosquito repellent chemicals effective on sweating skin, Research Project X-168, Report No. 4, Naval Med. Research Inst., June 8, 1945.

——, ——, and M. L. Hopwood: Summary of studies on new insect repellents, Research Project X-168, Report No. 8, Naval Med. Research Inst., Sept. 25, 1945.

——, ——, —— and L. M. Kozloff: The oxidation of 1,2,3,4, tetrahydro beta naphthol in relation to mosquito repellent action, Research Project X-168, Report No. 5, Naval Med. Research Inst., Aug. 1, 1945.

Poulton, E. B.: "The Colours of Animals," 2d ed. London, Paul, Trench, Trübner & Co., 1890.

Ralston, A. W., and J. P. Barrett: Insect repellent activity of fatty acid derivatives, Oil & Soap, 18(1), 89–91 (1941).

Richardson, C. H.: Advances in entomology, Ind. Eng. Chem., News Ed., 18(2), 64–72 (1940), 19(2), 77–88 (1941); 20(4), 241–256 (1942).

Ripley, L. B., B. K. Petty, and P. W. Van Heerden: Studies on gustatory reactions and feeding of wattle bagworm, with special reference to dusted foliage, Union S. Africa, Dep. Agr., Forestry Sci. Bull. 148, 1–27 (1936).

——, ——, and ——: Further studies on gustatory reactions of the wattle bagworm (Acanthopsyche junodi Heyl), Ibid., 205, 1–20 (1939).

Roark, R. C.: "An Index of Patented Mothproofing Materials," U.S. Dep. Agr., 1931, 125 pp.

——: "A Second Index of Mothproofing Materials," Ibid., 1933, 109 pp.

——: "A Third Index of Mothproofing Materials," Ibid., 1936, 104 pp.

Roth, L. M.: The odoriferous glands in the Tenebrionidae, Ann. Entomol. Soc. Am., 38(1), 77–87 (1945).

Rudolfs, W.: Effect of chemicals upon the behavior of mosquitoes, New Jersey Agr. Exp. Sta., Bull. 496, 1–24 (1930).

Schulz, P.: Die Nackengabel der Papilionidraupen, Zool. Jahrb. Anat., 32(2), 181–244 (1911).

Searls, E. M., and R. Daehnert: A synthetic fly repellent proves effective, Wisconsin Exp. Sta. Bull., 451, 43–44 (1941).

Shaw, A. O., R. C. Smith, F. W. Atkeson, H. C. Fryer, A. R. Borgmann, and F. J. Holmes: Tests of fly repellents of known ingredients and of selected commercial sprays on dairy cattle, J. Econ. Entomol., 36(1), 23–32 (1943).

Shaw, F. R.: Bee poisoning: A review of the more important literature, Ibid., 34(1), 16–21 (1941).

Shepard, H. H.: "The Chemistry and Toxicology of Insecticides," Minneapolis, Burgess Pub. Co., 1941.

Smirnow, D. A.: Ueber den Bau und die Bedeutung der Stinkdrüsen von Aromia moschata L., Trav. soc. imp. nat. St. Petersbourg, Sect. zool. physiol., 40(3), 1–15 (1911).

Stäger, R.: Neue Versuche über die Einwirkung von Duftstoffen und Pflanzendüften auf Ameisen, Mitt. schweizer. entomol. Ges., 15(13), 567–584 (1933); Biol. Abstr., 10(1), 1515 (1936).

Sweetman, H. L.: "The Biological Control of Insects," Ithaca, Comstock Pub. Co., 1936.

——, and A. I. Bourne: The protective value of asphalt laminated paper against certain insects, J. Econ. Entomol., 37(5), 605–609 (1944).

Thomssen, E. G., and M. H. Doner: Attractants and repellents for insects, Soap Sanit. Chemicals, 18(4), 97–99, 105, 106; 18(5), 95, 96, 105 (1942).

Turnbull, W. H.: History of the use of bee repellents in orchard sprays in the Okanagan valley of British Columbia, *Proc. Entomol. Soc. Brit. Columbia* **42,** 7–8 (1945).

U.S. Dep. Agr.: Insects in relation to national defense: Flies, Circ. 8, 1941.

——: Insects in relation to national defense: Devices for insect control, Circ. 20, 1941.

Van Antwerpen, F. J.: Chemical repellents, *Ind. Eng. Chem. Ind. Ed.*, **33**(12), 1514–1518 (1941).

Volkonsky, M.: A new method of protecting crops from Acrididae, *Compt. rend. soc. biol.*, **125,** 417–418 (1937).

Wain, R. L.: Secretion of salicylaldehyde by the larvae of the brassy willow beetle (*Phyllodecta vitellinae* L.), *Ann. Rep. Agr. Hort. Research Sta. (Bristol)*, 1943, pp. 108–110.

Wallace, P.: Chemical repellents to bark beetle breeding. *Connecticut Agr. Exp. Sta. Bull.* 445, 374–375 (1941).

Wardle, R. A., and P. Buckle: "The Principles of Insect Control," Manchester, University Press, 1923.

Weber, H.: "Biologie der Hemipteren," Berlin, Springer, 1930.

Wegener, M.: Die biologische Bedeutung de Nackengabel der Papilionidraupen. *Biol. Zentr.*, **43**(3), 292–301 (1923).

Whitcomb, W. D.: The carrot rust fly, *Mass. Exp. Sta., Bull.* 352, 1–36 (1938).

Whitten, R. R.: Toxic and repellent sprays for the control of elm bark beetles, *U.S. Dep. Agr., Circ.* 647, 1–12 (1942).

Williams, O. L.: Wood preference tests, "Termites and Termite Control," University California Press, 1934, pp. 532–533.

Woglum, R. S., and H. C. Lewis: Whitewash to control potato leafhopper on citrus, *J. Econ. Entomol.*, **33**(1), 83–85 (1940).

Wolcott, G. N.: Phenol as a termite repellent, *Science*, **101**(2626), 444 (1945).

Yothers, M. A., and F. W. Carlson: Repellency of pyrethrum extract and other materials to full-grown codling moth larvae, *J. Econ. Entomol.*, **37**(5), 617–620 (1944).

von Zehmen, H.: Versuche über Abschreckmittel für Nonnenraupen im Jahre 1938, *Anz. Schädlingskunde*, **18**(12), 137–139 (1942).

Chapter 9

Chemical Basis of Taste and Olfaction

With the unprecedented impetus that has been imparted to the study of chemoreception by demands for more efficient repellents there has developed coincidentally a realization of the extent of ignorance surrounding the relation between the chemistry of stimulative substances and their attractive or repellent properties. That this phase of insect physiology is of unquestioned practical moment has already been indicated in the previous chapter. Certainly it would appear to be of such a fundamental nature as to have claimed the attention of many physiologists. Paradoxical as it may appear, few workers have struggled with the problem.

If we continue to employ the words tastes (or taste-substances) and odors to designate two kinds of chemical stimuli that affect profoundly the behavior of insects, remembering of course that considerable indecision concerning distinctions between the two persists but that despite their anthropomorphic flavor and broad, flexible, rather indecisive connotation they remain universally convenient tools, then the problem may be presented succinctly as *What is the chemical basis of taste and odor?*

In reviewing the foundation of incontrovertible facts upon which our problem rests let us use the compound noun, taste-substances, as before to connote chemical stimuli which act by contact as liquids or solutions, and the word, odor, to signify chemical stimuli in a gaseous state (cf. pp. 12, 253). Then these are the facts that have been established: (1) insects possess the ability to taste, (2) they possess the ability to perceive odors, (3) the gustatory sense is distinct from the olfactory sense, (4) the gustatory end organs are located on portions of the antennae, palpi, hypopharynx, epipharynx, labrum, and legs, (5) the olfactory end organs are located on the antennae and mouthparts, (6) taste and odorous substances are either acceptable or unacceptable. Beyond this we enter a field of greater uncertainty and through lack of basic facts must draw by analogy from the records of vertebrate physiology.

233

Taste-Substances. The categories of substances that are known at this writing to be taste-substances with reference to insects are limited in number. Chief among them are sugars and electrolytes. Sugars are generally acceptable though some may appear tasteless or be indifferently received. Bees, ants, flies, and butterflies accept most sugars; codling moth larvae respond positively to brown sugar, corn sirup, sorbitol, sucrose, d-fructose, and glycerol; tent caterpillars and others accept sucrose, glucose, and fructose. Electrolytes are, on the whole, unacceptable. Most caterpillars refuse HCl; wattle bagworms refuse 1 per cent solutions of tartaric and oxalic acids. At low concentrations, however, some acids may be acceptable. Codling moth larvae accept, for example, malic acid; wattle bagworms accept acetic, citric, malic, and tannic acids. While salts are usually unacceptable, some may, at certain concentrations, be accepted. Such are bismuth carbonate and 3 per cent alum, two salts acceptable to wattle bagworms.

Aside from sugars and electrolytes few other substances are known with certainty to stimulate gustatory end organs of insects. For the most part, the evidence regarding the stimulating power of certain other crystalloid organic compounds known to taste sweet to man is questionable. Considerably more work is required to establish the rôle of glycerol, saccharine, glycols, and alcohols as acceptable taste-substances. On the other hand, Dethier and Chadwick (1947) recently showed that rejection thresholds may be obtained for aliphatic alcohols when these are applied to the tarsal chemoreceptors of blowflies. Formerly the barriers to the testing of organic compounds of this sort had been either their insolubility in nontoxic solvents, their own toxicity, or their odors. The latter tended to confuse the issue by stimulating olfactory or common chemical nerve endings. These difficulties were resolved by removing the olfactory organs prior to testing. Tests were then conducted as described for threshold determinations.

Alcohols. Dethier and Chadwick found that a high degree of correlation exists between the mean concentrations of alcohols at rejection threshold of the tarsal chemoreceptors of blowflies and such properties as boiling point, vapor pressure, molecular surface, molecular moment, water-oil distribution coefficients, standard free energies, and activity coefficients. This leaves little doubt that the experiments dealt primarily with the receptor cells rather than with some other link in the complex which intervenes between presentation of the stimulus and the response upon which the measurements depended.

If the pri-n-alcohols alone are considered, the results approximate to a Traube series (Traube, 1891), in which the effective concentrations are related as $1:3^{-1}:3^{-2} \ldots 3^{1-n}$, where n is the number of carbon atoms in the chain. Actual values from the data on blowflies are $1:3.49^{-1}:2.98^{-2}:2.60^{-3}:3.26^{-4}:3.92^{-5} \ldots :3.51^{-7}$. The discrepancies from a constant ratio are significant but of about the same order of magnitude as those noted with a variety of other material. The review on cell permeability by Davson and Danielli (1943) may be consulted for reference.

Conformity to this type of pattern indicates that stimulation of the tarsal receptors is not dependent on osmotic pressure nor on rate of molecular diffusion in solution, which decrease as the series is ascended. The relationship between stimulatory efficiency (1/threshold) and vapor pressure also is inverse. Simple correspondence between stimulating power and molecular weight or number of CH_2-groups is refuted by the results with several isomeric alcohols tested since the mean concentrations at threshold are significantly different for all pairs with the same number of carbon atoms except for two (i-butyl and n-butyl, i-amyl and n-amyl).

From the facts determined the logical inference is that some property shared by the entire series and varying with molecular structure in an orderly manner is concerned intimately with the stimulatory process. This property is dependent upon surface energy relationships, but it would be fruitless at this time to attempt to define the exact mechanism of its operation. We know nothing of the molecular structure of the receptor surface, and while the results indicate that its properties are similar to those of other cell membranes throughout the animal and plant kingdoms, it is not even certain that penetration of the surface is essential to stimulation.

Sugars. As Wigglesworth (1939) has pointed out, different species of insects taste, as sweet, different numbers of sugars; and many substances which appear sweet to man are tasteless to various insects. von Frisch's work with the honey bee has demonstrated that pentoses (arabinose, xylose, rhamnose), sugar alcohols (mannitol, dulcitol, sorbitol, erythritol), and many mono-, di-, and trisaccharides are tasteless. So little comparative work has been essayed in surveying the stimulative powers of different sweet substances for numbers of species that it is impossible at this time even to attempt to correlate the acceptability or the modality of sweet substances with structural formulae or with physical properties. On the other hand, recent work has thrown some light on the probable mechanism of action of electrolytes.

Acids. With respect to human beings, the sour taste is identified with the hydrogen ion. Insects also are sensitive to differences in hydrogen-ion concentration. As is the case with leafhoppers (cf. p. 250) pH is frequently of great importance in regulating feeding behavior. That sensitivity to the hydrogen ion on the part of insects may correspond to the sour taste by man is suggested by a similarity between the comparative stimulative efficiencies of hydrochloric and acetic acids in the two cases. Judged by rejection thresholds, acetic acid is considerably more effective as a stimulating agent than is hydrochloric acid at the same pH. For insects the ratio is from 20:1 to 25:1; for man, 28:1.

Inequalities of the stimulating efficiencies of different acids of equal pH indicate that the intrinsic pH is not the only variable factor involved in stimulation. It has been postulated that for proper stimulation an acid must combine with the taste-hair (in vertebrates) or even penetrate, albeit partially, to the interior of the taste-cell. This conception has led to general studies of the penetration of cells by acids and more especially with the aid of tissues containing natural indicator pigments (Crozier, 1916, 1916a, 1916b, 1922). Data from experiments dealing with penetration indicate that internal pH is an important variable. To borrow from Crozier's (1934) words, "The excitation of the sour taste may therefore be considered to depend upon the H ion, brought into the surface layer of the taste-cell by diffusion of the acid . . . , and there reacting with some specific taste-cell mechanism. The speed with which the acid penetrates determines the intensity of the excitation."

Just how closely these ideas may conform to the situation as it exists in insects is uncertain. The experiments of Chadwick and Dethier (1947) with the tarsal receptors of blowflies yield results which are in striking agreement with those of Crozier for penetration of essentially the same groups of acids into the mantle cells of *Chromodoris*.

With the saturated fatty acids from acetic through valeric the concentrations required for rejection decrease with increasing molecular weight. Formic acid is too stimulating on the basis of concentration alone to fit into its expected place at the beginning of this series, yet it falls into line if its greater degree of dissociation is taken into account. That is, when the several acids are compared in respect to pH at mean rejection threshold, rather than concentration, the order is: formic, acetic, propionic, *n*-butyric, *n*- and *i*-valeric. This result shows that the stimulating power of the acids apart from their capacity for producing hydrogen ions increases with increasing chain length. When, however, the pH at rejection threshold is plotted against number of carbon atoms or such properties as boiling point or oil-water distribution coefficients, the relationship is nonlinear on both arithmetic and logarithmic grids, and it is found that the increment in stimulative effect grows smaller as the series is ascended.

Basically similar results were obtained with the three dicarboxylic acids, oxalic, malonic, and succinic, which form an orderly series when compared in respect to pH at rejection threshold. Each had a lower pH at rejection than the corresponding monocarboxylic acid. This indicates that the anions or free molecules of the dicarboxylic series are relatively less stimulating.

As judged by the mean concentrations required for rejection, acetic acid is rendered considerably more stimulating by the introduction of one chlorine atom and still more so by three, but the greater stimulating power of the chloro-substituted acids is accounted for by their greater degree of dissociation which results in a higher concentration of hydrogen ions relative to the amount of acid dissolved. Actually, then, substitution of chlorine for hydrogen reduces the stimulative power of the anion or free acid, but this is overshadowed by the large increase in degree of dissociation.

The effect of substitution of —OH for —H was investigated in two series of acids: propionic-lactic-glyceric and succinic-malic-tartaric. In both series the introduction of a hydroxy group reduces the pH at rejection, and a further reduction occurs when a second hydroxy is substituted. Evidently the replacement of —H by —OH lowers the stimulating power of the anion or free acid, and this must be compensated by an increase in concentration of hydrogen ions in order for stimulation to occur.

Pyruvic was the only keto acid tested. It may be compared with propionic and lactic (= 2-hydroxypropanoic) which have the same number of carbon atoms. As noted above, substitution of a hydroxy group decreases the stimulating effect of the anion or unionized acid and requires a compensatory increase in concentration of hydrogen ions. The results with pyruvic acid indicate that the stimulative power is weakened even more by the keto grouping.

Fumaric and maleic acids were tested as examples of compounds containing a double bond and also as a check on the influence of stereoisomeric configuration. The succinic-malic-tartaric series may be used for comparison. The pH of fumaric (= *trans*-butenedioic) at rejection is almost identical with that of *d*-tartaric (2.19 and 2.18). Thus the effect of the double bond is equivalent, approximately, to the introduction of two hydroxy groups. Malic acid (= *cis*-butenedioic) is rejected only when a pH of 1.94 has been reached, but since it is more highly dissociated than fumaric the molar concentration at rejection is somewhat less. Since the *cis*-form is chemically the less stable of the two, this comparison also suggests that the limiting process in tarsal stimulation is not the chemical interaction between the acids and some cellular constituent.

The similarity of these results with Crozier's provides a strong reason for believing that penetration of the receptors is the limiting process in tarsal stimulation although it does not prove beyond doubt that penetration is an essential preliminary to sensory reception. The factors, presumably surface energy relationships, which favor penetration by many of the compounds in question would also favor their adsorption at a lipid-water or lipo-protein interface, and it is possible, in the absence of decisive evidence one way or the other, that simple accumulation of taste-substances in the cell membrane would result in excitation.

In cases where penetration by acids is known to occur there has been a great deal of discussion as to whether they enter the cell as ions or in the undissociated state, with perhaps the majority of writers upholding the latter alternative (cf. Davson and Danielli, 1943); but, as pointed out by Crozier (1916), it is difficult to reconcile this concept with stimulation by the mineral acids which are considered to be completely ionized in the concentrations required for sensory excitation. The results of Chadwick and Dethier with the monocarboxylic series also are more readily understood in terms of penetration or accumulation in the ionic form, for the order of

decreasing ionic concentrations at rejection threshold is that of increasing molecular weights while no such correspondence is found
when the series is arranged in order of decreasing concentrations of
free acid. Because only a single acid, formic, mars the sequence this
discrepancy would carry little weight if it were not for irregularities
revealed in several of the other series when they are arranged in
order of concentration of free acid. Thus with increasing chain
length the dicarboxylic acids show an increase in the amount of free
acid at the rejection threshold. Among the hydroxy substituted
acids the monocarboxylic series shows first a decrease in free acid
and then a rise as first one, then two, hydrogens are replaced, but
the dicarboxylic series yields a decrease in free acid with both steps
of hydroxy substitution. It is true that all these differences could be
reconciled by the assignment of appropriate arbitrary values for
the stimulating power of the several free acid molecules, but it
would seem more likely that the stimulating power changes in an
orderly manner with successive similar changes in molecular constitution. Also, if the acids are acting mainly in the unionized form,
it is difficult to see why the concentrations of free acid at rejection
threshold in the monocarboxylic group do not form a Traube series
as is the case with the corresponding alcohols. For these reasons
Chadwick and Dethier, pending the development of more direct
evidence, prefer the hypothesis that it is the anions, rather than the
free acid molecules, which have furnished the second major component of tarsal stimulation.

Whether this detail of interpretation is eventually confirmed or
disproved, it is quite apparent that the same general qualities determine the relative effectiveness of the individual acids in all the
series examined. As the affinity of the acid for water is increased, the
stimulating power of the anion or free acid is diminished. This is
true whether the structural change consists in removal of a CH_2-
group, substitution of $-Cl$, $-OH$, $=O$, or $-COOH$, replacement
of a $C-C$ linkage by a double bond, or a shift in stereoisomeric configuration. Thus it is clear that stimulation by the acids involves
processes very similar to those inferred previously for the series of
aliphatic alcohols, with the difference that in the case of the acids
the correlation between surface energy relationships and stimulatory
power is obscured by the high effectiveness of the hydrogen ion.
Since the tendency of the acids to dissociate commonly parallels

their hydrophile character, weakening of the stimulative power of the anion or free acid molecule is accomplished by an increase in the ease of production of hydrogen ions. The actual effect on the receptors depends then on the balance struck between these two opposing factors.

Salts. The comparative stimulative efficiencies of other ions for insects that have been tested and for various species of animals ap-

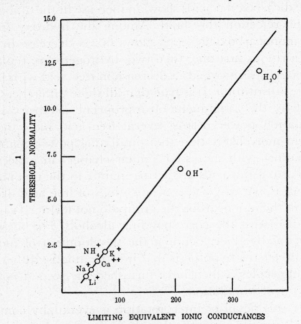

LIMITING EQUIVALENT IONIC CONDUCTANCES

FIG. 63. Graph illustrating the relationship between the stimulative efficiencies, for cecropia moth larvae, and the limiting equivalent ionic conductances in Mhos at 25° C. of certain ions. (Redrawn after Frings, *Biol. Bull.*, 1945.)

pear to be similar. Using larvae of *Platysamia cecropia* and adult roaches, Frings (1945 and 1946) determined the order of relative stimulative efficiencies as measured by reciprocals of the normalities of rejection limens (thresholds) for various cations, as chlorides, as: $NH_4^+ = K^+ > Ca^{++} > Na^+ > Li^+$. In all cases there is a linear relation between the stimulative efficiencies of these cations and their limiting equivalent ionic conductances (ionic mobilities). To Frings these relationships suggest a common mode of action and possibly a common modality of taste. Emphasizing the mere suggestiveness here involved he hypothecates, as did Eger (1937), only two taste modalities for insects—acceptable and unacceptable. Bitter, salt, and sour modalities may be merely isolated points on a continuous concentration spectrum Such a consideration is strengthened by the fact that the hydrogen ion, popularly responsible for sour taste, falls in line with the ions of salts when its stimulative efficiency is plotted against its limiting equivalent ionic conductance (Fig. 63)

Odors. The number and kinds of substances that may stimulate olfactory receptors are legion. Beyond the basic facts stated initially

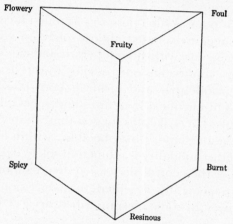

FIG. 64. Henning's olfactory prism.

our concepts of this aspect of chemoreception are illusive. For even a minimum of theoretical discussion it is necessary to analogize with human physiology.

Of the handful of attempts to classify odors, Henning's (1924) scheme, imperfect at best, emerges as the most convenient, Conceived as an olfactory prism this scheme represents each corner as one of six fundamental classes: flowery, fruity, spicy, resinous, burnt, and foul (Fig. 64). Intermediates are plotted in appropriate positions on the respective surfaces.

Modalities. Henning's hypothesis, forwarded to explain the manner of action of odorous compounds by relating certain aspects of structure with the class of odor that the compound emanates, is ingenious but, as Crozier (1934) pointed out, is based upon conceptions not always acceptable to the organic chemist. Moreover, "smell" is hardly a property of the substance alone; it is a property also of the receptor. Briefly, the hypothesis attempts to define odors in terms of the structure of the ring molecule. Substitution groups that are odor-generating are hydroxyl, aldehydic, ketonic, ester, and nitril, but the type of odor is determined chiefly by the total molecular configuration. Molecules responsible for a spicy odor have substitutes in a para-position; flowery groups have substitutes in meta- or ortho-positions; fruity groups are composed of molecules with branched side chains; resinous substances are inner ring compounds; burnt groups are characterized by simple rings; foul groups possess broken rings.

Evidence that odor as such possesses as much meaning for insects as for man is suggestive. Based almost solely on von Frisch's conditioning experiments with honey bees it indicates that the same similar odors are confused as one by some insects as are confused by man. Thus benzaldehyde and furfural, both possessing the same odor but different chemical structures and physical properties, are indistinguishable. Facts like this suggest that whatever is responsible for odor as we understand it may be alike, within limits, for insects and man. Whether modality is a function solely of properties of the end organs or of the central nervous system cannot be stated dogmatically. It is generally accepted that sensation results from a change of activity in some circumscribed area of the central nervous system.

Acceptance and Rejection. Attempts to relate repellency and attrahency to molecular content and configuration have been singularly unsuccessful. Scattered published reports illustrate this point. Speyer reported in 1920 that for house flies aromatic compounds, especially essential oils, are poor attractants as compared to aliphatic acids, aldehydes, and alcohols. Bunker and Hirschfelder (1925), attempting to relate repellency (to mosquitoes) to chemical structure, found indications that effectiveness is somehow related to the oxygen atom, especially $\equiv$C—O—H in alcohols, —C$\equiv$O$_2$ of esters, and $=$C$=$O of ketones regardless of whether the molecule is aliphatic or carbocyclic. Hydrocarbons are considerably improved by the introduction of oxygen. Thus the addition of OH to methyl benzoate and amyl benzoate changes them to methyl and amyl salicylate respectively, two good repellents.

The superiority of acids and esters over alcohols as codling moth attractants and of aromatic acids over aliphatic and simple dibasic acids (Eyer and Medler, 1940) does not appear to be a general phenomenon inasmuch as alcohols and aliphatic compounds are superior fly attractants. Nor does the reported increase of repellency from hydrocarbon to alcohol to ester (Moore, 1934) appear to be general either.

With regard to esters, Cook (1926) found that the addition of CH_2 groups to the alcohol side produces little effect on the compound as a fly attractant while addition of CH_2 to the acid side is attended by a profound increase in attractiveness. This coincides with the observations of Eyer, Medler and Linton (1937) and Eyer and Medler (1940) that from the point of view of codling moth attractants the best esters are those of the methyl series when the alcohol radicals are compared and those of the propyl series when the acid radicals are compared.

The only generalization that can be made is that iso compounds usually are more effective as excitants than are normal compounds. Cook has pointed out by way of explanation that iso compounds have the lower boiling points and that the difference in attraction from normal compounds increases as the boiling point increases.

There are as yet no indications that the difference between attrahency (acceptance) and repellency (rejection) is based upon the constituents or configuration of a molecule except as these two variables may modify physical properties. As there appears to be a critical concentration for every compound (provided it is an adequate stimulus in the first place) at which it acts as a repellent we may best speak of odorous substances in terms of their exciting power or stimulative efficiency. Gustatory and olfactory stimuli are alike in this respect in possessing an acceptance threshold and a rejection threshold (cf. pp. 163, 234). Stimulative efficiency appears to be more directly related to physical properties than to molecular configuration. This is most strikingly observed when homologs or sterioisomers are compared.

Specificity of repellents or, more correctly speaking, different stimulative efficiencies of a given substance for different insects may arise as a result of dissimilar properties of the external cuticular parts of the sensilla, unlike thresholds, or unlike patterns or properties of the central nervous systems. We do know from conditioning experiments that the reaction to any given odor, that is, whether the odor acts as an attractant or a repellent, is determined in the final analysis by the central nervous system. Increasing the intensity of an odor (i.e., the concentration of the stimulating chemical) so that the response of the animal may change from positive to negative, causes an increase in the frequency of impulses discharged by the receptor.

Intensity of Odor. How effective a substance may be in stimulating olfactory receptors, irrespective of whether it acts as an attractant or a repellent, seems to be determined by several factors associated with physical properties. These in turn are related to the properties of the end organ. Probably the first prerequisite for successful stimulation is the ability to penetrate the cuticular covering of end organs and probably also the sense cell membranes in sufficient concentration. The second is that the substance must be an adequate stimulus. With regard to the latter requirement it may be pointed out that certain gases such as oxygen, and liquids such as water, may amply satisfy the first requirement and fail to stimulate. Yet it is obvious that not even an adequate stimulus can stimulate unless it reaches the end organ. In view of these requirements such properties of a compound as boiling point, molecular weight, and solubility are important. Until more is learned concerning the chemicophysical properties of cuticle surmounting chemoreceptors, it is difficult to relate the two series (*re* cuticle cf. Hurst, 1941; Trim, 1941; Richards and Anderson, 1942; Beament, 1945; Wigglesworth, 1945, etc.).

Boiling Point. Stimulating effectiveness of odors decreases with increasing boiling point or, in other words, the more volatile a substance the better it stimulates, all other things being equal. Such a relationship is not entirely unexpected when it is recalled that the principal prerequisite of olfactory stimulation is volatility. Thus the most volatile compound may be more active initially, but its period of activity is of shorter duration. The work of Cook (1926) designed to determine the most attractive concentration over a given period of time of certain paraffin derivatives for flies is illustrative. By exposing each test solution till no more flies were attracted he found that during the first 24-hour period 7 per cent ethyl alcohol was the most attractive; during the second, 1 per cent iso-propyl; during the third, 0.25 per cent butyl. For the total length of time iso-propyl caught the greatest number of flies.

When the log of concentration is plotted against the log of boiling points, the result is a straight line, i.e., the most attractive concentration of an alcohol or ester of the paraffin series decreases with increase in boiling point (Figs. 65 and 66). Such a relation might be expected to hold for all odorous materials. In this particular case the log of optimum solution $= 13.78 \pm 7.1$ log boiling point, or the effective amount, decreases approximately 7 times as rapidly as log boiling point increases.

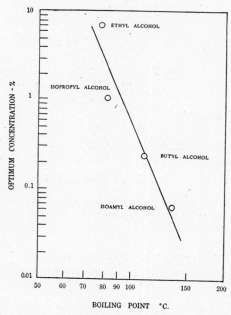

Fig. 65. Relation between boiling point and the most attractive concentration (as indicated by number of houseflies caught) of aliphatic alcohols. (Redrawn after Cook, *J. Agr. Research*, 1926.)

Molecular Weight. In general, intensity of odor increases with increasing molecular weight to a maximum, after which effectiveness declines (Crozier, 1934). This proposition appears to hold true for insects as well as for vertebrates. Higher members of a homologous series of compounds stimulate more effectively than lower members. In the case of houseflies, aliphatic acids, aldehydes, and alcohols are attractive provided the molecular weight exceeds 30 (Speyer, 1920). Amyl groups are especially attractive. Codling

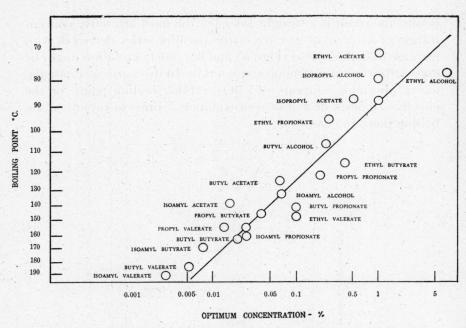

FIG. 66. Relation between boiling point and the most attractive concentration (as indicated by number of houseflies caught) of paraffin compounds. (Redrawn after Cook, *J. Agr. Research*, 1926.)

moths respond most readily to aromatic acids containing six carbon atoms, an arrangement which approximates the optimum molecular weight.

Solubility. From the human point of view most taste substances are water soluble, but as odors must be able to penetrate the mucous layer at the olfactory epithelium and the lipoid substance of the olfactory end organ as well, an effective olfactory compound must needs be both water and oil soluble. There are indications that oil solubility is the more important of the two properties. The intensity of action of an odor-excitant may then truly be conditioned by solubility. If solubility in nonpolar liquids or cell surface constituents is important, then the relation between intensity of odor and molecular weight is understandable since solubility increases with increasing molecular weight.

Even this meager fund of information is not available regarding insects. It is especially significant that the olfactory end organs of insects differ markedly from those of vertebrates in the possession of a cuticular covering of singular properties. Thus, hypotheses correlating the physical properties of substances odorous to man with intensity of action cannot be extended to insects.

REFERENCES

Adrian, E. D.: "The Basis of Sensation," New York, W. W. Norton & Co., 1928.

Arey, L. B., and W. J. Crozier: The sensory reactions of *Chiton*, *J. Exp. Zoöl.*, **29**(2), 157–260 (1919).

Beament, J. W. L.: The cuticular lipoids of insects, *J. Exp. Biol.*, **21**(3/4), 115–131 (1945).

Best, C. H., and N. B. Taylor: "The Physiological Basis of Medical Practice," 4th ed., Baltimore, The Williams and Wilkins Co., 1945.

Cameron, A. T.: The relative sweetness of various sweet compounds and of their mixtures, *Can. J. Research*, **23**(5), 139–166 (1945).

Chadwick, L. E., and V. G. Dethier: The relationship between chemical structure and the response of blowflies to tarsal stimulation by aliphatic acids. *J. Gen. Physiol.*, **30**(3), 255–262 (1947).

Cole, W. H.: Stimulation by the salts of the normal aliphatic acids in the rock barnacle *Balanus balanoides*, *Ibid.*, **15**(6), 611–620 (1932).

——, and J. B. Allison: Chemical stimulation by alcohols in the barnacle, the frog and planaria, *Ibid.*, **14**(1), 71–86 (1930).

——, and ——: Stimulation by mineral and fatty acids in the barnacle *Balanus balanoides*, *Ibid.*, **16**(6), 895–903 (1933).

——, and ——: Stimulation by the mineral acids, hydrochloric, sulfuric, and nitric, in the sunfish *Eupomotis*, *Ibid.*, **16**(4), 677–684 (1933a).

Cook, W. C.: The effectiveness of certain paraffin derivatives in attracting flies, *J. Agr. Research*, **32**(4), 347–358 (1926).

Crozier, W. J.: Cell penetration by acids, *J. Biol. Chem.*, **24**(3), 255–279 (1916).

——: Cell penetration by acids. II. Further observations on the blue pigment of *Chromodoris zebra*, *Ibid.*, **26**(1), 217–223 (1916a).

——: Cell penetration by acids. III. Data on some additional acids, *Ibid.*, **26**(1), 225–230 (1916b).

——: Cell penetration by acids. VI. The chloroacetic acids. *J. Gen. Physiol.*, **5**(1), 65–79 (1922).

——: Chemoreception. "A Handbook of General Experimental Psychology," Worcester, Mass., Clark University Press, 1934, pp. 987–1036.

Davson, H., and J. F. Danielli: "The Permeability of Natural Membranes," New York, The Macmillan Company, 1943.

Dethier, V. G.: Gustation and olfaction in lepidopterous larvae, *Biol. Bull.*, **72**(1), 7–23 (1937).

——: Taste thresholds in lepidopterous larvae, *Ibid.*, **76**(3), 325–329 (1939).

——, and L. E. Chadwick: Rejection thresholds of the blowfly for a series of aliphatic alcohols, *J. Gen. Physiol.*, **30**(3), 247–253 (1947).

Eyer, J. R., and J. T. Medler: Attractiveness to codling moth of substances related to those elaborated by heterofermentative bacteria in baits, *J. Econ. Entomol.*, **33**(6), 933–940 (1940)

——, ——, and H. P. Linton: Analysis of attrahent factors in fermenting baits used for codling moth, *Ibid.*, **30**(5), 750–756 (1937).

Frings, H.: Gustatory rejection thresholds for the larvae of the cecropia moth, *Samia cecropia* (Linn.), *Biol. Bull.*, **88**(1), 37–43 (1945).

——: Gustatory thresholds for sucrose and electrolytes for the cockroach, *Periplaneta americana* (Linn.), *J. Exp. Zoöl.*, **102**, 23–50 (1946).

von Frisch, K.: Vergleichende Physiologie des Geruchs- Geschmackssinnes, *Handb. norm. path. Physiol.*, **2**, 203–239 (1926).

Henning, H.: "Der Geruch," Leipzig, Barth, 1924, 434 pp.

Hopkins, A. E.: Chemical stimulation by salts in the oyster *Ostrea virginica*, *J. Exp. Zoöl.*, **61**, 13–28 (1932).

Howlett, F. M.: The effect of oil of citronella on two species of *Dacus*, *Trans. Entomol. Soc. London*, **1912**, 412–418.

Hurst, H.: Insect cuticle as an asymmetrical membrane, *Nature*, **147**(3726), 388–389 (1941).

Minnich, D. E.: The chemical senses of insects, *Quart. Rev. Biol.*, **4**(1), 100–112 (1929).

Moore, W.: Esters as repellents, *J. N. Y. Entomol. Soc.*, **42**(2), 185–192 (1934).

Osterhout, W. J. V.: Physiological studies of single plant cells, *Biol. Rev.*, **6**(4), 269–411 (1931).

Parker, G. H.: "Smell, Taste, and Allied Senses in the Vertebrates," Philadelphia, J. B. Lippincott Co., 1922.

——, and E. M. Stabler: On certain distinctions between taste and smell, *Am. J. Physiol.*, **32**(4), 230–240 (1913).

Parry, E. J.: "The Chemistry of Essential Oils and Artificial Perfumes," 4th ed., London, Scott, Greenwood and Son, 1931, v. 2, Chap. 2.

Richards, A. G., and T. F. Anderson: Electron microscope studies of insect cuticle, with a discussion of the application of electron optics to this problem, *J. Morph.*, **71**(1), 135–184 (1942).

Severin, H. H. P., and H. C. Severin: Relative attractiveness of vegetable, animal, and petroleum oils for the Mediterranean fruit fly (*Ceratitis capitata* Wied.), *J. N. Y. Entomol. Soc.*, **22**(3), 240–248 (1914).

Speyer, E. R.: Notes on chemotropism in the house-fly, *Ann. Applied Biol.*, **7**(1), 124–140 (1920).

Taylor, N. W.: A physico-chemical theory of sweet and bitter taste excitation based on the properties of the plasma membrane, *Protoplasma*, **4**(1), 1–17 (1928).

Traube, J.: Ueber die Capillaritätsconstanten organischer Stoffe in wässerigen Lösungen, *Ann. Chem.*, **265**, 27–55 (1891).

Trim, A. R. H.: The protein of insect cuticle, *Nature*, **147** (3717), 115–116 (1941).

Webb, J. E., and R. A. Green: On the penetration of insecticides through the insect cuticle, *J. Exp. Biol.* **22**(1/2), 8–20 (1945).

Wigglesworth, V. B.: "The Principles of Insect Physiology," New York, E. P. Dutton & Co., 1939.

——: Transpiration through the cuticle of insects, *J. Exp. Biol.*, **21**(3/4), 97–114 (1945).

Chapter 10
Evolution of Feeding Preferences

Interesting by-products of studies relating to naturally occurring attractants and repellents are insights into (1) the evolution of feeding habits in different groups, and (2) contemporary changes and fluxes in feeding habits. Basic knowledge acquired from studies of this nature may be applied with profit to the development of insect-resistant plant varieties.

Mechanism of Choice. The existence of feeding preferences is a logical and well-established fact. Such preferences are diverse and varied, restricted or unlimited. As an insect can select only from among those hosts and plants available to it, preferences are subject to seasonal, ecological, and geographical limitations. On the part of the insect they may vary also with species, race, sex, developmental stage, physiological age, or even with individuals. They are most in evidence among parasites, where they may more properly be referred to as host preferences, and among phytophagous insects. The ability to exercise a choice presupposes a means of recognition. Parasites are guided by host odors serving as attractants (see Chapter 2). Phytophagous insects are guided by plant odors. Here morphological features in the form of physical repellents (hair, pubescence, etc.) and repellent tastes further modify the choice (see Chapter 8). Some insects even exhibit marked preferences in the selection of drinking water. In instances of this sort they are guided by a water-perception sense (Hertz, 1934) and an olfactory appreciation of volatile organic compounds present in the water (Butler, 1940). Parasites usually show a choice ranging from generic or broader down to specific. There are even instances known where particular host tissues are selected. The encyrtid *Diversinervus smithi* deposits eggs only in the fused ganglion of the central nervous system of certain species of lecaniine scale insects (Flanders, 1944). Plant-feeders frequently show such fine preferences involving distinctions between races, strains, certain anatomical parts, or tissues. Whereas host and plant choice is based primarily on odor and secondarily on external physical characters, plant tissue choice appears to be based entirely on the sense of taste, on the ability to dis-

tinguish differences in pH. Tissue choice attains its highest
development in the plant lice and leaf hoppers. Coördinating the
observations of Bennett (1934 and 1937) and Bennett and Esau
(1936) that the virus of curly top of beets is almost exclusively a
phloem disease and that the beet leaf hopper normally feeds on
phloem, Fife and Frampton (1936) presented a hypothesis attempting to explain the method of tissue selection by these insects. The
hypothesis is based on the existence of a pH gradient between cell
contents of phloem and of parenchyma. Measurements of the acidity of cell sap with a potentiometer and a quinhydrone microelec-
trode showed that phloem sap is more alkaline than parenchyma
sap. This indicates the presence of a pH gradient (Fig. 67). It was
hypothesized that aphids were guided to the phloem by the pH

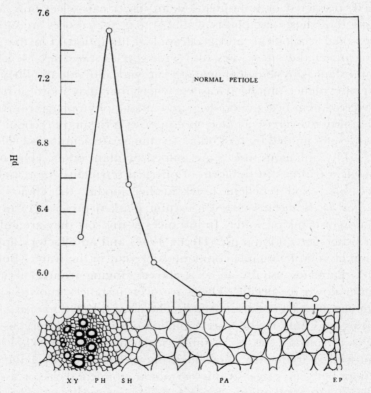

FIG. 67. Diagram illustrating the pH gradient in a normal sugar-beet petiole:
(XY) xylem, (PH) phloem, (SH) bundle sheath, (PA) parenchyma, (EP) epidermis. (Redrawn after Fife and Frampton, *J. Agr. Research*, 1936.)

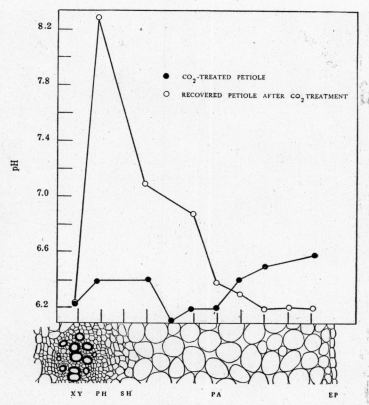

FIG. 68. Diagram illustrating the pH gradient in a sugar-beet petiole treated with CO_2: (XY) xylem, (PH) phloem, (SH) bundle sheath, (PA) parenchyma, (EP) epidermis. (Redrawn after Fife and Frampton, *J. Agr. Research*, 1936.)

gradient. When the mouth-parts penetrate the parenchyma, the difference in acidity between cell sap and the saliva gives rise to an unfavorable reaction causing withdrawal or change of locale of the mouth-parts.

Additional experiments supported this hypothesis. Leaf hoppers were placed in a feeding cage where they could pierce a membrane and feed on sugar drops adjusted to two different pH values (8.5 and 5.0). They were observed to explore the acid drops with their mouth-parts and move to the alkaline ones where they finally settled down to feed. Further evidence was procured by subjecting beets to carbon dioxide in high concentration to modify the pH gradient in normal tissue. This tended to equalize the phloem and

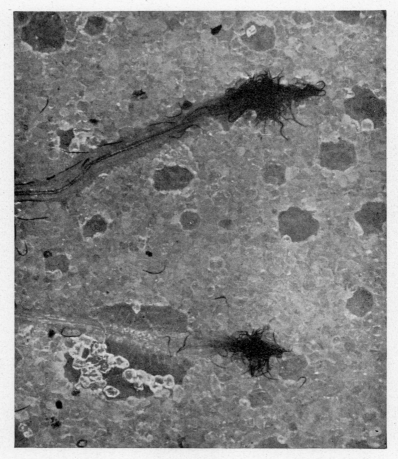

FIG. 69. Nematode larvae attracted to host-plant root tips by exudations of chemicals which are dispersed in soil water. (Courtesy, M. B. Linford.)

parenchyma pH (Fig. 68). Leaf hoppers feeding on these treated beets fed indiscriminately on different tissues.

Studies of this nature of six different species of *Empoasca* (Smith and Poos, 1931) revealed that five species, *E. maligna, E. abrupta, E. filamenta, E. bifurcata,* and *E. erigeron* selected mesophyl tissue of more mature leaves. *E. fabae* preferred phloem tissue.

The sense of taste also has been revealed as important in regulating the food selection of subterranean root-feeders. Investigation was first directed toward larvae of nematode worms. Information gathered from this source, together with data obtained from studies of wireworms, presents a unified picture of selection that undoubtedly applies to the majority of subterranean root-feeding insects.

Linford (1939) demonstrated with eelworm larvae in sand that worms are attracted to fresh and decaying cut leaves, stems, and roots by chemical compounds. Hosts are recognized by exudations of chemicals from the roots as these exudations are dispersed in soil water (Fig. 69). Worms apparently locate a root by traveling from a region of soil water of low concentration to regions of higher concentration. Gadd and Loos (1941), testing the comparative attractiveness of roots in moist sand, found that worms may be attracted even to roots which subsequently prove so distasteful as to repel them. These authors postulated that eelworms are guided to living roots in preference to decaying ones because of a higher concentration of attractant within the former.

Wireworms behave in a similar manner. Whereas they are unable to orient to air-borne odors of favored food, they do react to water-borne substances. Extracts of food plants initiate two distinct responses, a biting response when applied experimentally to filter paper, and an aggregation response in sand. Thorpe, et al. (1945) found that biting of baited filter paper was elicited by sugars, fats, polypeptides, and to some extent by tannin and polyhydric alcohols. Aggregation is initiated by asparagine, glutamine (a constituent of potato, one of the food plants), and related compounds such as aspartic and malic acids. Amines of lower fatty acids are active while the acids themselves are inactive. Whereas sugars elicit both responses, other substances which have been tested elicit but one. Thus asparagine does not cause biting nor do triolein or casein cause aggregation. From these observations Thorpe, et al., concluded that there must exist two types of chemoreceptors, stimulation of the one leading to aggregation and of the other to biting. The sensitivity of the former is comparable to that of olfactory receptors; the sensitivity of the latter, to gustatory receptors. Apparently, wireworms locate their food by a klino-kinetic reaction. They wander into a region of active substance, and their behavior remains the same till they stray out of this region. Then by increased turning they wander back until eventually a substance is encountered which initiates a biting reaction.

Monophagy, Oligophagy, and Polyphagy. Recognition of the intimate relation between plant chemistry and the choice of food by phytophagous insects has made possible an understanding of the feeding habits of this group not possible with parasites where host odor chemistry is still largely a mystery. It was customary to divide the group into three categories on the basis of the diversity of the food eaten. As originally proposed, the terms monophagous, oligophagous, and polyphagous were intended to designate respectively, insects with a single food plant, those with a few very definite food plants, and those which feed indiscriminately on many species.

A truly monophagous insect is exceptionally rare. Said to be monophagous are the beech aphis, *Coccus fagi*, which is restricted to beech and avoids even the copper variety, and the Mexican boll weevil which feeds on *Gossypium*. Recent experiments indicate, however, that the latter will develop completely on *Hibiscus syriacus* L. The only well-known truly monophagous lepidopterous larvae are those of the paw paw butterfly and *Lycaena hypophlaeas*. The latter feeds on *Rumex acetosella* L., but subsequent experiments may prove that even this insect will accept other plants. Numerous insects are monophagous, not because they will accept only one plant, but because other acceptable plants possess different geographical ranges. *Danais plexippus*, which in some parts of its range subsists upon a single species of *Asclepias*, will, if presented with them, eat a wide variety of tropical Asclepiadaceae and Apocynaceae. It will also readily eat *Acerates* and some Cactaceae. All of these plants emit the same odor.

Insects classed as oligophagous select a few very definite plants because (1) these often unrelated plants contain the same attracting chemical, or (2) the insect is conditioned to several chemicals. In this respect *Papilio ajax* is a fine example of an oligophagous insect. *P. ajax* larvae eat some species of *Citrus*, *Ruta*, and *Zanthoxylum* because all contain methylnonylketone. Coriander, carrot, caraway, and anise are eaten because the larvae are conditioned also to the different types of chemicals found therein. It has been shown that the larvae do not confuse all eight attracting chemicals but rather recognize each as separate and distinct. It does not follow, however, that a polyphagous larva, as for example *Platysamia cecropia* L., with 60-odd food plants recognizes 60 different chemicals or confuses the 60 as one. Instead, it would seem that such a larva is conditioned to no particular chemical and eats any plant not actually repellent.

In the light of these considerations it is proposed that the term monophagous be restricted to an insect attracted by a single chemical or a group of closely related chemicals confused as one; oligophagous, an insect attracted by a few specific and different chemicals and capable of distinguishing one from the other; polyphagous, an insect which, on the whole, is not particularly attracted by any chemical but accepts as food any plant not repellent by reason of chemical or physical characteristics.

Genetic Basis of Selection. As pointed out earlier (p. 162) the nature of the genetically directed response to an odorous substance determines it as an attractant or a repellent. Thus the attraction to specific odors of monophagous and oligophagous species is genetically directed. Polyphagy also has a genetic basis. There is considerable evidence to confirm this assertion that many preferences are genetically fixed: (1) Thorpe and Jones (1937) have proved that the attractiveness of *Ephestia* host odor to the parasite *Habrobracon* is germinally fixed; (2) a more or less regular inheritance of food habits has been suggested in hybrids between species of *Basilarchia;* (3) races and strains differing from each other by feeding habits only have been found to be genetically distinct and can be crossed only with difficulty.

Nevertheless, genetically determined preferences are not so dominant that they may not be influenced by internal and external factors. The expression of genes is considerably influenced by external factors of temperature, humidity, etc., and internal physiological factors the most influential of which is hormonal. The behavior of *Pimpla ruficollis*, a parasite of the pine shoot moth (*Rhyacionia buoliana* Schiff.) toward an essential oil has been shown by Thorpe and Caudle (1938) to be an association of this last sort. Emerging from its host long before the next generation of host larvae is available for oviposition, the parasite forsakes pine trees for Umbelliferae upon which it then commences to feed. At this time its ovaries are small and oil of *Pinus sylvestris* is repellent. Later, in the third or fourth week of life, the ovaries enlarge, oil of pine becomes attractive, and the females return to pine where hosts are now available. Another example of physiological conditions affecting chemotactic responses is seen in the behavior of the parasite *Coccophagus caridei*. Unmated females oviposit in parasitized mealy bugs, but mated females restrict their attention to lecaniine scales. Physiological changes also are presumed to account for differences of feeding habits of a given insect in different stages and a seasonal alternation of feeding habits as exemplified by certain aphids which choose a winter host, usually woody, and one or more summer hosts (*Aphis bakeri* Cowen winters on apple and summers on clover).

Conditioning. Auxiliary and superficial preferences operate in the absence of, or obscure, germinally fixed preferences. These result from conditioning or, more correctly as Thorpe points out, from "becoming aware" (habituation) of a new odor as part of a favorable environment, a behavior also observed in ants by Forel and Fielde. An insect may be naturally or experimentally conditioned to odors. There are numerous examples of experimental conditioning including that of bees by von Frisch and blowflies by Frings. That this has its limitations, however, is seen by the fact that bees may be conditioned to many aromatic odors but never to foul odors such as skatole and asafetida. Then, too, the conditioning is of a transient nature. It disappears after about one week. In some species it may be readily reëstablished; in others it is extinct. In the case of *Habrobracon* Thorpe has shown that there is a preimaginal conditioning of developing parasites to the odor of an experimental host but that the preferences derived from this host conditioning are in no manner so powerful as the germinally fixed preference for the normal host odor. He has shown also (1939) that other insects permit of preimaginal conditioning. *Drosophila* larvae reared on a food scented with peppermint produced adults still partial to peppermint odor. This was true even when last instar larvae were thoroughly washed, proving that influences operative only in the larval stage can modify adult behavior. That they significantly modify oviposition responses has been demonstrated by Cushing (1941) in experiments with the fungus-inhabiting fly, *Drosophila guttifera*. This induced behavior, however, is of short duration. Crombie (1944) found that blowflies first exposed to the odor of menthol (normally repellent) in the larval stage became so habituated that the adults tested in an olfactometer were indifferent to the odor. This habituation, however, was only of a few days duration and failed when tested against an odor concentration approaching saturation.

There is no irrefutable evidence that such conditioning is inherited. All apparent inheritance or strengthening of conditioned preferences may be explained by selection. As Thorpe pointed out, any factor tending to split a population into groups will tend to aid in the establishment of new variants just like geographical barriers, and no Lamarckian Theory is necessary.

Hopkins Host Principle and Biological Races. Since the enunciation of the Hopkins Host Principle, which states in essence that an oligophagous or polyphagous insect prefers to oviposit on the food upon which it fed as a larva, there has been much conflicting evidence both for and against the hypothesis. Generally speaking the principle appears to hold. As pointed out above, conditioning of larvae to particular foods may be retained through pupation to the adult stage at which time the insect shows preference for that food upon which it fed as a larva. It then oviposits on the preferred food and tends to perpetuate the behavior. Adaptation of this sort to a new environment tends to split a species into feeding groups. This condition in itself has a tendency to prevent crossbreeding for, as Huxley (1942) explained, it provides a nonhereditary barrier which may serve as the first stage in evolutionary divergence. This is the Baldwin and Lloyd Morgan Principle of Organic Selection: (1) An organism becomes adapted to an ecological niche by behavior and consequent nonheritable modifications. (2) Mutations for the kind of structural modifications suitable to that particular mode of life have a better opportunity for being selected. This ecobiotic isolation is similar to geographic isolation.

All of this makes for biological and physiological races. Thorpe defined them as divisions of a species into groups usually isolated to some extent by food preferences, occurring in the same locality, and showing definite differences in biology (habits, means of oviposition, etc.) but with corresponding structural differences either few and inconstant or completely absent. Such races are genetic. They are said to be the products of mutant genes.

Evolution of Feeding Habits. It is now generally accepted that polyphagy represents a more primitive state than monophagy. There is ample evidence to indicate, however, that polyphagy might also have arisen secondarily from monophagy. The monophagous state has been derived through mutation from monophagous races within a polyphagous species. Apparently conflicting evidence supporting both schools of thought as to primitive feeding habits lends credence to the existence of an alternation of monophagy and polyphagy during evolution. From the chemical side there is new evidence that changes in both directions are in progress, at least in the Lepidoptera.

Lepidopterous larvae are by no means static as regards feeding habits. Changes have occurred in the evolutionary history of a species and may even now be taking place. Such changes may be explained as (1) reversion to an acceptable plant not previously available, (2) a gradual adaptive conditioning to a new chemical, or (3) a mutation in hereditary instinct.

Reversions are undoubtedly responsible for most of the changes in feeding habits listed in the literature. The most spectacular instance was that of *Papilio zelicaon* Luc. in California. This species usually found feeding on Umbelliferae recently reverted to *Citrus*. Brues (1924) surmised that this represented a reversion, and proof of that fact is the occurrence of *P. ajax* and other Umbelliferae-feeders on rutaceous plants, especially since extensive feeding experiments indicate that the Umbelliferae-feeding habit probably gradually evolved from a Rutaceae-feeding habit (Dethier, 1941). Similar cases of reversion are well known. Particularly striking is the shift of some Australian Papilios to introduced *Citrus* from domestic and endemic plants (Waterhouse and Lyell, 1914).

The existence of slow gradual shifts to new food plants is difficult to establish although it has been accomplished experimentally. The hypothesized evolution of feeding habits in the genus *Papilio* best illustrates this type of change.

The 400 or so species in the genus *Papilio* may be grouped into four categories: (1) *Aristolochia*-feeders; (2) species whose feeding habits are varied; (3) Umbelliferae-feeders; and (4) Rutaceae-feeders. The first two categories have no direct bearing on this study, but investigation into the feeding habits of the latter two groups sheds light on the apparent trends of evolution of feeding habits in this section of the genus.

Only 11 species of *Papilio* are now known to eat Umbelliferae. These are *P. ajax* L. and its races, *P. bairdi* Edw. and its races, *P. zelicaon* Luc., *P. indra* Reak., *P. machaon* L. and its races, *P. hospiton* Gen., *P. alexanor* Esp., *P. demoleus* L., *P. ophidecephalus*, *P. constantinus*, and *P. paeon* Bdv. Doubtless the list may be lengthened when our knowledge of the feeding habits of the genus has widened. It seems strange that in a large cosmopolitan group of more or less standard feeding habits a few species should have become rabidly addicted to Umbelliferae. A careful study of food plant lists, plus a consideration of plant chemistry, suggests a way in which this habit may have become initiated. The fact that Umbelliferae-feeders as exemplified by *P. ajax* will still eat species of *Ruta* suggests that they probably formerly belonged to the Rutaceae-feeders.

To understand the transition from Rutaceae to Umbelliferae, it is necessary first to consider the changes that have taken place in Rutaceae-feeders. Rutaceous plants on the basis of their essential oils (Table 5) may be divided into four groups: those with lemon- or orangelike odor containing citral (e.g., *Citrus*), those with a rue odor containing methylnonylketone (e.g., *Ruta*), those containing both oils in varying proportion (e.g., *Zanthoxylum*), and those of mint or camphor type. Most of the evidence points toward species of *Citrus* as being the original host plants of Rutaceae-feeders. The transition from *Citrus*-feeders to *Ruta*-feeders may have been gradual, that is, *Citrus* to *Zanthoxylum* to *Ruta*. Then the transition from Rutaceae-feeders may have been effected by means of *Dictamnus Fraxinella* which contains methyl chavicol and anethole, or *Pelea madagascarica* which contains anethole. It seems probable, however, that the change to such plants as *D. Fraxinella* and *P. madagascarica* proceeded by way of intermediates, *Zanthoxylum* and *Ruta*, and not directly from *Citrus*. The proposed line of evolution would then be *Citrus* (and plants containing citral) to *Zanthoxylum* (or other plants containing citral and methylnonylketone in varying proportions) to *Ruta* (or plants containing methylnonylketone) to *D. Fraxinella* (or similar species). Or *Ruta* may have been omitted from the system. *P. cresphontes*, for example, feeds upon: *Citrus*, *Ptelea trifoliata* L., *Zanthoxylum americanum* Mill., *Z. Clava-Herculis* L., *Z. ajuda*, *Dictamnus Fraxinella* Pers., *Nyssa sylvatica* Marsh., *Persea carolinensis* Nees., *Populus pyramidalis* Ait., *Piper peltatum* L., *P. umbellatum* L., *P. aduncum*, and *P. mollicomum* Kunth. In Cuba it also feeds on species of *Triphasia*, *Zanthoxylum coriaceum* A. Rich, *Atalantia ceylonica* Am. Oliver, *Feronia limonia* Swingle, and *Fortunella crassifolia* Swingle. These plants contain citral and possess an odor of lemon or orange. *Feronia elephantum*, which smelled faintly of rue, was nibbled only slightly. Some species such as *P. andraemon*, *P. cresphontes*, *P. machaon*, and *P. ajax* represent various steps in the change from one feeding habit to a new and different one. The acceptance as food of plants from other families (species of *Piper*, for example) may be explained in part by their chemical composition (cf. Table 5).

In just what direction *P. hospiton* and *P. alexanor* are headed is difficult to determine in the face of such a paucity of facts. Whether or not they eat any plants other than *Ferula* and *Seseli* is not known. If they are attracted by the characteristic odor of sulfides in these plants, they may be induced to feed on *Allium*.

This is one way in which ecobiotic isolation facilitating operation of the Baldwin Lloyd Morgan Principle may be brought about.

Radical departures from a fixed feeding habit not explainable on the basis of the foregoing propositions are infrequent. Supposedly they arise as outright mutations in the genetic constitution governing the instinctive attraction to specific compounds. The acquisition of a leguminous food plant by essentially crucifer-feeding Pieridae is assumed to be a mutation of this sort (Brues, 1924). These shifts in feeding habits are sudden, complete, constant, and independent of any morphological or physiological changes in the insect. In these respects they resemble structural changes usually associated with gene mutations.

Conclusion. Probably the most interesting contribution of the study of chemical attractants and repellents to the biology of feeding is the clarification of the conception of the oligophagous state. The fact that some larvae are restricted to plants of a certain family or genus and other larvae to a different group is due almost entirely to the chemical nature of the plants concerned and in no way to the taxonomic relationship. The tendency of certain chemicals to be restricted to a particular plant family or genus is not unusual, and this, coupled with the fact that insects are attracted to the chemicals, actually not to the plant, does much to explain the so-called "botanical instinct."

REFERENCES

Abbott, C. E.: The physiology of insect senses, *Entomologica Am.*, **16**(4), 225–280 (1937).

Ball, E. D.: The food plants of the leafhoppers, *Ann. Entomol. Soc. Am.*, **25**(3), 497–501 (1932).

Bennett, C. W.: Plant-tissue relations of the sugar-beet curly-top virus, *J. Agr. Research*, **48**(8), 665–701 (1934).

——: Correlation between movement of curly top virus and translocation of food in tobacco and sugar beet, *Ibid.*, **54**(7), 479–502 (1937).

———, and K. Esau: Further studies on the relation of the curly top virus to plant tissues, *Ibid.*, **53**(8), 595–620 (1936).

Brues, C. T.: The selection of food-plants by insects with special reference to lepidopterous larvae, *Am. Naturalist*, **54,** 313–332 (1920).

———: The specificity of food-plants in the evolution of phytophagous insects, *Ibid.*, **58,** 127–144 (1924).

Butler, C. G.: The choice of drinking water by the honeybee, *J. Exp. Biol.*, **17**(3), 253–261 (1940).

Calvert, P. P.: The geographical distribution of insects and the age and area hypothesis of Dr. J. C. Willis, *Am. Naturalist*, **57**(650), 218–229 (1923).

Cheo, Ming-tsang: A preliminary list of the insects and arachnids injurious to economic plants in China, *Peking Nat. Hist. Bull.*, **11**(2), 119–127, (1936).

Coad, B. R.: Feeding habits of the boll weevil on plants other than cotton, *J. Agr. Research*, **2**(3): 235–245 (1914).

Comstock, J. A.: "Butterflies of California," Los Angeles, 1927.

Craighead, F. C.: Hopkins host-selection principle as related to certain cerambycid beetles, *J. Agr. Research*, **22**(4), 189–220 (1921).

Crombie, A. C.: On the measurement and modification of the olfactory responses of blowflies, *J. Exp. Biol.*, **20**(2), 159–166 (1944).

Cushing, J. E.: An experiment on olfactory conditioning in Drosophila guttifera, *Proc. Nat. Acad. Sci.*, **27**(11), 496–499 (1941).

Czapek, F.: "Biochemie der Pflanzen," Jena, Gustav Fischer, 1921, Vols. 1–3.

De Bach, P.: Environmental contamination by an insect parasite and the effect on host selection, *Ann. Entomol. Soc. Am.*, **37**(1), 70–74 (1944).

Dethier, V. G.: Gustation and olfaction in lepidopterous larvae, *Biol. Bull.*, **72**(1), 7–23 (1937).

———: Taste thresholds in lepidopterous larvae, *Ibid.*, **76**(3), 325–329 (1939).

———: Chemical factors determining the choice of food plants by Papilio larvae, *Am. Naturalist*, **75,** 61–73 (1941).

———: The function of the antennal receptors in lepidopterous larvae, *Biol. Bull.*, **80**(3), 403–414 (1941).

———: Testing attractants and repellents, "Laboratory Procedures in Studies of the Chemical Control of Insects," Washington, D. C., Am. Assoc. Adv. Sci., 1943, pp. 167–172.

Engler, A., and K. Prantl: "Die natürlichen Pflanzenfamilien," Leipzig, W. Engelmann, 1894–1898.

Fielde, A. M.: Power of recognition among ants, *Biol. Bull.*, **7**(5), 227–250 (1904).

Fife, J. M., and V. L. Frampton: The pH gradient extending from the phloem into the parenchyma of the sugar beet and its relation to the feeding behavior of *Eutettix tenellus*, *J. Agr. Research*, **53**(8), 581–593 (1936).

Finnemore, H.: "The Essential Oils," New York, D. van Nostrand Co., 1926.

Flanders, S. E.: Olfactory responses of parasitic Hymenoptera in relation to their mass production, *J. Econ. Entomol.*, **37**(5), 711–712 (1944).

Folsom, J. W.: A chemotropometer, *Ibid.*, **24**(4), 827–833 (1931).

———, and R. A. Wardle: "Entomology with Special Reference to its Ecological Aspects," Philadelphia, The Blakiston Company, 1934.

Forel, A.: "Le Monde Social des Fourmis du Globe," Genève, Forel, Kundig, 1921–1923, 5 vols.

Frings, H.: The loci of the olfactory end-organs in the blowfly, *Cynomyia cadaverina* Desvoidy, *J. Exp. Zoöl.*, **88**(1), 65–93 (1941).

Frost, S. W.: A study of the leaf mining Diptera of North America, *Agr. Exp. Sta.*, *Cornell Univ. Mem.* 78, 1924.

Fulton, B. B.: Physiological variation in the snowy tree-cricket, *Oecanthus niveus* De Geer, *Ann. Entomol. Soc. Am.*, **18**(3), 363–383 (1925).

Funkhauser, W. D.: Biology of the membracidae of the Cayuga Lake Basin, *Agr. Exp. Sta., Cornell Univ. Mem.* 11, 1917.

Gadd, C. H., and C. A. Loos: Host specialization of *Anguillulina pratensis* (De Man). I. Attractiveness of roots, *Ann. Applied Biol.*, **28**(4), 372–381 (1941).

Gahan, A. B.: A list of the phytophagous Chalcidoidea with descriptions of two new species, *Proc. Entomol. Soc. Washington*, **24**, 33–58 (1922).

Gildemeister, E., and F. Hoffmann: "The Volatile Oils," Milwaukee, Pharm. Review Pub. Co., 1900.

Götz, B.: Beiträge zur Analyse des Verhaltens von Schmetterlingsraupen beim Aufsuchen des Futters und des Verpuppungsplatzes, *Z. vergl. Physiol.*, **23**(3), 429–503 (1936).

Hamlin, J. C.: An inquiry into the stability and restriction of feeding habits of certain cactus insects, *Ann. Entomol. Soc. Am.*, **25**(1), 89–120 (1932).

Harrison, J. W. H.: On the inheritance of food habits in the hybrids between the geometrid moths (*P. pomonaria*, Hb. and *P. isabellae*, Harrison), *Proc. Univ. Durham Phil. Soc.*, **7**, 194–201 (1920).

Hayes, W. P.: Biological races of insects and their bearing on host plant resistance, *Entomol. News*, **46**(1), 20–23 (1935).

Hertz, Mathilde: Eine Bienendressur auf Wasser, *Z. vergl. Physiol.* **21**(3), 463–467 (1934).

Huxley, J.: "Evolution: The Modern Synthesis," London, Geo. Allen & Unwin, Ltd., 1942.

Imms, A. D.: "Recent Advances in Entomology," Philadelphia, The Blakiston Company, 1931.

Kono, H., and T. Sawamoto: On the foods of the larvae of the butterflies of Hokkaido, *Kontyu*, **12**(4), 140–145 (1938).

Langford, G. S., M. H. Muma, and E. N. Cory: Attractiveness of certain plant constituents to the Japanese beetle, *J. Econ. Entomol.*, **36**(2), 248–252 (1943).

Larson, A. O.: The host-selection principle as applied to *Bruchus quadrimaculatus* Fab., *Ann. Entomol. Soc. Am.*, **20**(1), 37–78 (1927).

Lima, A. M. Da Costa: "Terceiro catalogo dos insectos que vivem nas plantas do Brazil," Rio de Janeiro, Ministerio da Agricultura, Dep. Nac. da Prod. veg. Escola Nac. de Agron., 1936.

Linford, M. B.: Attractiveness of roots and excised shoot tissues to certain nematodes, *Proc. Helminth. Soc. Washington*, **6**(1), 11–18 (1939).

Lovell, J. H.: Origin of oligotrophy in bees, *Entomol. News*, **25** (7), 314–321 (1914).

McIndoo, N. E.: The olfactory sense of lepidopterous larvae, *Ann. Entomol. Soc. Am.*, **12**(2), 65–84 (1919).

——: The relative attractiveness of certain solanaceous plants to the Colorado potato beetle, *Leptinotarsa decemlineata* Say, *Proc. Entomol. Soc. Washington*, **37**(2), 36–42 (1935).

McNair, J. B.: Some properties of plant substances in relation to climate of habitat —volatile oils, saponins, cyanogenetic glucosides, and carbohydrates, *Am. J. Bot.*, **19**(2), 168–193 (1932).

Meissner, O.: Monophagie und Polyphagie, *Int. Entomol. Zeit.*, **20**, 130–132 (1926).

*Mell, R.: Biologie und Systematik der südchinesichen Sphingiden, 1922.

Meyrick, R.: The hereditary choice of food plants in the Lepidoptera and its evolutionary significance, *Nature*, **119**, 388 (1927).

Mosher, F. H.: Food plants of the gipsy moth in America, *U.S. Dep. Agr., Bull.* 250, 1–39 (1915).

Painter, R. H.: The food of insects and its relation to resistance of plants to insect attack, *Am. Naturalist*, **70**(731), 547–566 (1936).

Parry, E. J.: "The Chemistry of Essential Oils and Artificial Perfumes," 3d ed., London, Scott, Greenwood and Son, 1918, 2 vols.

Pavlov, I. P.: "Conditioned Reflexes: An Investigation of the Activity of the Cerebral Cortex," Oxford University Press, 1927.

Robertson, C.: Origin of oligotrophy in bees, *Entomol. News*, **25**, 67–73 (1914).

Salt, G.: Experimental studies in insect parasitism. III. Host selection, *Proc. Roy. Soc. London*, 117(B805), 413–435 (1935).

——: The sense used by *Trichogramma* to distinguish between parasitized and unparasitized hosts, *Ibid.*, **122**(B826), 57–75 (1937).

Seitz, A.: "The Macrolepidoptera of the World," Stuttgart, Fritz Lehman, 1906–1932.

Sladden, D. E.: Transference of induced food-habit from parent to offspring. Part I. *Proc. Roy. Soc. London*, **114**(B790), 441–449 (1934); II. *Ibid.*, **119**(B812), 31–46 (1935).

——, and H. R. Hewer: Transference of induced food-habit from parent to offspring. III. *Ibid.*, **126**(B842), 30–44 (1938).

Smith, F. F., and F. W. Poos: The feeding habits of some leaf hoppers of the genus *Empoasca, J. Agr. Research*, **43**(3), 267–285 (1931).

Sweetman, H. L.: Notes on insects inhabiting the roots of weeds, *Ann. Entomol. Soc. Am.*, **21**(4), 594–600 (1928).

Takahashi, R.: Monophagy and polyphagy among phytophagous insects: An evolutionary consideration, *Kontyu*, **12**(4), 130–139 (1938).

Thorpe, W. H.: Biological races in *Hyponomeuta padella* (L.), *J. Linn. Soc. Zoöl.*, **36**, 621–634 (1929).

——: Biological races in insects and allied groups, *Biol. Rev.*, **5**(3), 177–212 (1930).

——: I. Biological races in insects and their significance in evolution, *Ann. Applied Biol.*, **18**(3), 406–414 (1931).

——: Further observations on biological races in *Hyponomeuta padella* (L.), *J. Linn. Soc. Zoöl.*, **37**(254), 489–492 (1931a).

——: Further experiments on olfactory conditioning in a parasitic insect. The nature of the conditioning process, *Proc. Roy. Soc. London*, **126**(B844), 370–397 (1938).

——: Further studies on pre-imaginal olfactory conditioning in insects, *Ibid.*, **127**(B848), 424–433 (1939).

——: "Ecology and the Future Systematics" in "The New Systematics," ed. by J. S. Huxley, Oxford University Press, 1940.

——: Types of learning in insects and other arthropods, *Brit. J. Psychology Gen. Sect.*, **33**(4), 220–234; **34**(1), 20–31; **34**(2), 66–76 (1943–1944).

——, and H. B. Caudle: A study of the olfactory responses of insect parasites to the food plant of their host, *Parasitology*, **30**(4), 523–528 (1938).

——, A. C. Crombie, R. Hill, and J. H. Darrah: The food finding of wire worms (*Agriotes* spp.), *Nature*, **155**(3924), 46–47 (1945).

——, and F. G. W. Jones: Olfactory conditioning in a parasitic insect and its relation to the problem of host selection, *Proc. Roy. Soc. London*, **124**(B834), 56–81 (1937).

Valentine, J. M.: The olfactory sense of the adult meal-worm beetle *Tenebrio molitor* (Linn.), *J. Exp. Zoöl.*, **58**, 165–228 (1931).

Varley, G. C.: On the search for hosts and the egg distribution of some chalcid parasites of the knapweed gall-fly, *Parasitology*, **33**(1), 47–66 (1941).

Wardle, R. A., and P. Buckle: "Principles of Insect Control," London, Manchester Univ. Press, 1923.

Waterhouse, G. A.: "The Biology and Taxonomy of the Australasian Butterflies," *Rep. 23d Meet. Austrl. and New Zealand Assoc. Adv. Sci.*, **1937**, pp. 101–133.

——, and G. Lyell: "The Butterflies of Australia," Sydney, Angus and Robertson, Ltd., 1914.

Wehmer, C.: "Die Pflanzenstoffe," Jena, G. Fischer, 1929–1935.

Weiss, H. B.: The similarity of insect food habit types on the Atlantic and Western Arctic coasts of America, *Am. Naturalist*, **60**(666), 102–104 (1926).

Wigglesworth, V. B.: The sensory physiology of the human louse *Pediculus humanus corporis* De Geer (Anoplura), *Parasitology*, **33**(1), 67–109 (1941).

Williston, S. W.: What is a species? *Am. Naturalist*, **42**(495), 184–194 (1908).

Author Index

A

Abbott, C. E., 21, 35, 111, 135, 160, 164, 168, 260
Acree, F., 36
Adrian, E. D., 247
Ahlgren, H. L., 73
Ahmad, T., 34, 35
Alessandrini, G., 131, 135
Alexander, C. C., 194, 195, 197
Allen, N., 94, 102
Allison, J. B., 247
Allison, V. C., 150–152, 168
Allman, S. L., 213, 226
Anderson, A. L., 166, 168
Anderson, T. F., 244, 248
Arey, L. B., 247
Atkeson, F. W., 227, 231
Atkinson, D. J., 72

B

Back, R. C., 36, 121, 132, 137, 183, 195, 199
Baker, A. C., 121, 134, 135, 192, 195, 197
Baker, F. E., 213, 227
Baker, W. C., 98, 102
Ball, E. D., 260
Bard, P., 108, 135
Bare, C. O., 35, 181, 197
Barnes, D. F., 194, 198
Barrett, J. P., 223, 231
Barritt, J., 228
Barrows, W. M., 3, 10, 95, 142–144, 168
Barth, G., 124, 135
Barth, R., 21, 23, 35
Bartlett, B. R., 119, 135, 181, 198
Beament, J. W. L., 244, 247
Belschner, H. G., 228
Belzung, E., 72
Benjamin, D. M., 119, 135
Bennett, C. W., 250, 260
Bentley, E. W., 15, 35
Benton, A. F., 154, 168
Best, C. H., 247
von Bichowsky, F. R., 154, 168
Bickley, W. E., 8, 11
Bishopp, F. C., 112, 138, 229, 230
Blagoveshchensky, D. I., 226
Blodgett, W. E., 215, 228
Blunk, H., 225, 226
Bobb, M. L., 91, 92, 100, 172, 173, 178, 180, 181, 192–194, 196–198

Bohorquez, R., 118, 135, 181, 198
Borgmann, A. R., 227, 231
Bourne, A. I., 225, 231
Bowe, E. E., 223, 224, 229
Boyce, A. M., 119, 135, 181, 198
Bradley, G. H., 135
Brandt, H., 10
Brauns, A., 130, 135
Bregetova, N. G., 226
Brighenti, D., 135
Brinley, F. J., 213, 226
Brody, A. L., 219, 220, 227
Brown, H. E., 174, 198, 200
Browne, C. A., 135
Brues, C. T., 258, 260, 261
Brunetti, C., 130, 135
Bua, G., 118, 120, 135, 181, 198
Buchanan, R. E., 100
Buchanan, W. D., 97, 100
Buckle, P., 11, 203, 232, 264
von Buddenbrock, W., 12, 35
Bunker, C. W. O., 227, 242
Burdette, R. C., 174, 198
Burgess, E. D., 8, 10, 172, 179, 186, 198
Butler, C. G., 57, 58, 74, 261
Buxton, P. A., 130, 135, 227

C

Calvert, P. P., 261
Cameron, A. T., 247
Cameron, T. W. M., 227
Carl, A. L., 72
Carlson, F. W., 194, 195, 197, 211, 232
Carter, H. F., 135
Caudle, H. B., 99, 102, 255, 263
Chadwick, L. E., 234, 236, 238, 239, 247
Chamberlin, J. C., 195, 198
Champion, H. G., 34, 35
Champlain, A. B., 100
Chaplet, A., 168
Cheo, Ming-tsang, 261
Chesnut, V. K., 65, 66, 74
Chisholm, R. D., 177, 198, 199
Christensen, C., 124, 137
Christophers, R., 216, 227
Clark, C. O., 227
Coad, B. R., 261
Cole, A. C., 183, 198
Cole, W. H., 247
Collins, C. W., 21, 23–25, 35
Comstock, J. A., 261
Conant, J. B., 16, 35

Cook, W. C., 243–247
Corbett, G. H., 115, 136
Cory, E. N., 73, 149, 168, 174, 175, 177, 186, 198–200, 262
Cragg, J. B., 112, 113, 133, 136
Craig, R., 57, 73, 112, 137, 149, 151, 169
Craighead, F. C., 98, 100, 261
Craigie, J. H., 124, 136
Crescitelli, F., 6, 7, 10
Crombie, A. C., 29, 35, 157–159, 168, 256, 261, 263
Crozier, W. J., 157, 168, 236, 238, 241, 245, 247
Crumb, S. E., 40, 69, 73, 117, 132, 136, 181, 199
Cuscianna, N., 101
Cushing, J. E., 256, 261
Czapek, F., 72, 261

D

Daehnert, R., 222, 231
Danielli, J. F., 235, 238, 247
Darrah, J. H., 263
Davidson, R. H., 110, 132, 136
Davson, H., 235, 238, 247
Dean, G. A., 2, 10
Dean, R. W., 119, 136, 181, 198
DeBach, P., 261
Dekay, H. G., 74
Delange, R., 59, 72
DeLong, D. M., 132, 136
Denham, C. S., 205, 227
Deonier, C. C., 222, 227
Dethier, V. G., 40, 41, 58, 71, 72, 111, 154, 157, 160, 166, 168, 173, 177, 180, 198, 234, 236, 238, 239, 247, 258, 261
Dewitz, J., 3, 10
Dickins, G. R., 21–23, 35
Dietrich, H., 96, 101
Dietz, H. F., 213, 228
Ditman, L. P., 149, 168, 174, 198
Doner, M. H., 231
Downes, W., 227
Dumont, C., 59, 72

E

Eagleson, C., 148, 168
Eger, H., 166, 169, 173, 177, 178, 180, 241
Eidmann, H., 21, 35
Ellison, L. O., 222, 229
Elsworth, F. F., 228
Engler, A., 261
Enzmann, E. V., 95, 101, 137, 177, 199

Esau, K., 250, 261
Evans, A. C., 112, 136
Eyer, J. R., 81, 83–90, 93, 101, 121, 132, 198, 243, 247

F

Fabre, J. H., 35, 125, 136
Fenton, F. A., 221, 228
Fielde, A. M., 256, 261
Fife, J. M., 250, 251, 261
Findlay, G. M., 211, 228
Finnemore, H., 72, 261
Fisher, R. A., 221, 227
Flanders, S. E., 206, 227, 249, 261
Fleming, W. E., 8, 10, 172, 177, 179, 186, 198, 213, 227
Flügge, G., 9, 10
Folsom, J. W., 28, 36, 65, 66, 72, 114, 147, 163, 164, 169, 203, 261
Forel, A., 256, 261
Fraenkel, G. S., 4, 10, 12, 36
Frampton, V. L., 250, 251, 261
Fredericq, L., 205, 227
Freeborn, S. B., 227
Freiling, H. H., 22, 36
French, M. H., 211, 228
Freney, M. R., 28, 29, 36, 112, 113, 136
Frings, H., 156, 166, 225, 227, 240, 248, 256, 261
von Frisch, K., 12, 21, 36, 156, 166, 169, 235, 242, 248, 256
Frost, S. W., 2, 10, 67, 72, 91–93, 96, 101, 262
Fryer, H. C., 221, 227, 231
Fuller, M. E., 29, 36, 113, 136
Fulmer, E. I., 100
Fulton, B. B., 203, 227, 262
Funkhauser, W. D., 262

G

Gadd, C. H., 253, 262
Gaddum, E. W., 174, 198
Gahan, A. B., 262
Garb, G., 205, 227
Garcia, F., 82, 83, 101
Gerjovich, H. J., 215, 227, 228, 231
Gildemeister, E., 73, 262
Gillett, J. D., 11
Gilmore, J. U., 40, 69, 73, 181, 198
Gimingham, C. T., 212, 227
Goetsch, W., 227
Gortner, R. A., 49, 73, 136
Götz, B., 21, 36, 262
Graenicher, S., 136

Granett, P. J., 228
Grant, G. H., 212, 230
Green, R. A., 248
Greshoff, M., 54, 73
Grüger, R., 227
Gunn, D. L., 4, 10, 12, 14, 36
Guy, H. G., 213, 228

H

Haberlandt, G., 47, 73
Haeussler, G. J., 6
Haller, H. L., 24, 36, 199
Hamlin, J. C., 262
Hammond, G. H., 225, 228
Hampton, F. A., 73
Hancock, J. L., 29, 36
Hansberry, R., 73, 210
Harrison, J. W. H., 262
Harrow, B., 75, 76, 101, 104, 105, 136
Hartley, R. S., 228
Hartung, E., 28, 36
Hartwell, R. A., 36
Hassan, A., 115, 136
Hayes, T. H., 136
Hayes, W. P., 262
Hayward, K. J., 120, 136
Hazard, F. O., 225, 228
Headlee, T. J., 132, 136, 182, 198
van Heerden, P. W., 71, 74, 202, 203, 231
Henning, H., 241, 248
Henschel, J., 36, 111, 136
Hepburn, G. A., 24, 38, 95, 102, 112, 113,
    120, 136, 145, 147, 148, 169, 177,
    178, 199
Heriot, A. D., 199
Hertz, M., 249, 262
Hettche, H. O., 101
Hewer, H. R., 263
Hewitt, C. G., 3, 10
Hey, D. H., 203, 230
Heymons, R., 136
Hill, R., 263
Hilton, W. A., 169
Hindmarsh, W. L., 228
Hines, H. J., 118, 138
Hinman, E. H., 131, 137
Hirschfelder, A. D., 227, 242
Hitti, P. K., 10
Hobson, R. P., 29, 36, 112, 137, 174
Hodson, A. C., 119, 120, 135, 137, 181,
    187, 190, 198, 199, 220, 222, 228
Hoffmann, F., 73, 262
Hogg, P. G., 73
Holde, D., 116, 137
Holden, J. R., 211, 228

Hollande, A. C., 205, 228
Hollowell, E. A., 203, 229
Holman, H. J., 228
Holmes, F. J., 231
Homp, R., 5, 10
Hooker, W. J., 73
Hopkins, A. E., 248
Hopwood, M. L., 227, 231
Hornby, H. E., 211, 228
Hoskins, W. M., 19, 38, 57, 73, 112, 117,
    121, 133, 137, 139, 149, 151, 153,
    160, 163–165, 169, 170, 174, 199,
    222, 229
Howell, D. E., 211, 228
Howlett, F. M., 10, 13, 19, 24, 36, 111,
    115, 137, 183, 199, 248
Huckett, H. C., 213, 228
Huffaker, C. B., 36, 121, 132, 137, 183,
    195, 199
Hull, J. B., 219, 228
Humphreys, W. J., 151, 169
Hundertmark, A., 6, 10
Hurst, H., 244, 248
Husain, M. A., 117
Hutner, S. H., 95, 101, 137, 177, 199
Huxley, J., 257, 262

I

Imms, A. D., 3, 10, 26, 27, 36, 117, 137,
    262
Ingle, L., 149, 150, 169
Ingram, A., 135

J

Jachowski, L. A., 215, 217, 228, 230, 231
Jahn, T. L., 6, 7, 10
James, H. C., 202, 228
Jannke, P. J., 71, 73
Jarvis, H., 118, 137
Jefferson, R. N., 91, 100, 172, 173, 178,
    180, 198
Jenkins, J. R. W., 213, 228
Jewett, H. H., 203, 229
Johnson, H. W., 203, 229
Johnson, J. P., 213
Jones, F. G. W., 27, 38, 148, 170, 255, 263
Jones, H. A., 173, 199, 200

K

Kaplan, H. M., 95, 101, 137, 177, 199
Katz, S. H., 150–152, 168
Kellog, V. L., 21, 36
Kemper, H., 229

Kennedy, J. S., 15, 37
Keys, A., 109, 138
Kilgore, L. B., 223, 229
Kirk, H. B., 100
Knipling, E. F., 219, 229
Koblitsky, L., 198, 199
Kofoid, C. A., 223, 224, 229
Kono, H., 262
Kozloff, L. M., 231

L

Laake, E. W., 112, 138, 222, 229, 230
Laing, J., 27, 37
Lamborn, W. A., 131, 137
Langford, G. S., 73, 175, 177, 179, 184–186, 199, 200, 262
Lantz, A. E., 98, 101
Larson, A. O., 203, 229, 262
Latossek, T., 227
Latter, O. H., 229
Lau, K. H., 114, 139
Lawson, F. R., 195, 198
Lea, A., 206, 229
Leach, J. G., 124, 137
Leblois, A., 73
Lees, A. D., 15, 37
Lehman, R. S., 25, 37, 146, 147, 169
von Lengerken, H., 136
Lesser, M. A., 229
Lewis, H. C., 213, 229, 232
Lima, A. M. Da Costa, 262
Lindquist, A. W., 219, 229
Linford, M. B., 252, 253, 262
Link, K. P., 73, 74
Linsley, E. G., 34, 37
Linton, H. P., 82, 101, 243, 247
Lipp, J. W., 29, 37, 213, 225, 229, 230
Loeffler, E. S., 133, 174, 199, 222, 229
Loos, C. A., 253, 262
Lovell, J. H., 262
Luftensteiner, H., 101
Lyell, G., 258, 264
Lyon, S. C., 40, 69, 117, 136, 181, 199

M

McColloch, J. W., 29, 37, 69, 73
McCoy, E. E., 7
McCulloch, R. N., 229
McDaniel, E. I., 213, 229
MacFie, J. W. S., 135
McGovran, E. R., 222, 229
McIndoo, N. E., 3, 10, 25, 37, 40, 65, 73, 115, 117, 121, 137, 143, 144–147, 169, 262

McNair, J. B., 262
McNay, C. G., 229
McPhail, M., 116, 118, 119, 135, 137, 181, 194, 197, 199
Madden, A. H., 219, 229
Maines, W. W., 8, 10, 172, 179, 186, 198
Manee, A. H., 34, 37
Mansbridge, G. H., 229
Marchal, P., 205, 230
Marchand, W., 109, 137
Marlowe, R. H., 213, 230
Marquez, V. M., 137
Marshall, J., 148, 166, 169
Martin, H., 202, 230
Martini, E., 10
Matheson, R., 131, 137
Matsurevich, I. K., 71, 73
Maxwell-Lefroy, H., 2, 10
Medler, J. T., 82–90, 101, 121, 132, 173, 177, 180, 198, 243, 247
Meissner, O., 262
Mell, R., 21, 23, 37, 262
Mesnil, L., 213, 230
Metzger, F. W., 35, 37, 177, 200, 212, 213, 230
Meyrick, R., 263
Mickelson, O., 109, 138
Middleton, W., 99, 101
Milam, J., 40, 69, 73, 171, 181, 186, 198, 200
Minaeff, M. G., 230
Minnich, D. E., 12, 37, 166, 169, 248
Miwa, Y., 194, 199
Monchadsky, A. S., 226
Mönnig, H. O., 138, 212, 230
Moore, 220, 222, 230
Moore, R. H., 40, 69, 73
Moore, W., 223, 230, 243, 247
Morgan, A. C., 40, 69, 73, 181, 199
Morgan, E., 212, 230
Moriyama, T., 194, 199, 202, 230
Mosher, F. H., 73, 263
Moskvin, I. A., 219, 230
Mulhearn, C. R., 230
Muma, M. H., 73, 115, 138, 175, 199, 262
Mumford, E. P., 203, 230
Munger, F., 174, 199
Murr, L., 21, 37, 110, 138

N

Nelson, E. K., 72
Nelson, O. A., 94, 102
Newman, L. J., 119, 138
Nicholas, J. E., 19, 38, 92, 102, 178, 184, 200

Nicholson, G. W., 37
Nolte, M. C. A., 112, 113, 136, 138, 206, 229
Noyes, W. M., 183, 199

O

O'Connor, B. A., 119, 138
Offhaus, K., 227
Orr, L. W., 124, 137
Oshima, M., 207, 230
Osterhout, W. J. V., 248

P

Pagden, H. T., 29, 37
Painter, R. H., 203, 230, 263
Parker, G. H., 12, 37, 248
Parker, J. R., 2, 10
Parman, D. C., 112, 113, 138, 229, 230
Parry, E. J., 44, 45, 73, 248, 263
Pavlov, I. P., 263
Pavlovskii, E. N., 230
Pearson, A. M., 230
Peffly, R. L., 132, 136
Perkins, F. A., 118, 138
Person, H. L., 96, 97, 101
Peterson, A., 2, 6, 11, 40, 53, 73, 81, 92, 101, 183, 194, 199
Petty, B. K., 71, 74, 202, 203, 231
Pierpont, R. L., 213, 230
Pijoan, M., 215–217, 228, 230, 231
Pincus, G., 157, 168
Plateau, F., 6, 11
Plummer, C. C., 135, 197
Plunkett, C. R., 17, 37, 107
Poos, F. W., 252, 263
Potts, S. F., 21, 23, 25, 35, 36
Poulton, E. R., 205, 231
Power, F. B., 65, 66, 74
Prantl, K., 261
Prüffer, J., 21, 37

R

Ralston, A. W., 223, 231
Ramage, G. R., 112, 113, 133, 136
Ramsay, J. A., 57, 58, 74
Raucourt, M., 40, 70, 74
Reed, M. R., 19, 37, 95, 102, 161–163, 169, 187, 189, 199
Rex, E. G., 182, 183, 199
Rheineck, A. E., 71, 74
Rhin, A. E., 74
Rhodes, H., 81, 101
Ribbands, C., 110, 138
Richards, A. G., 244, 248
Richardson, C. H., 26, 29, 37, 117, 138, 230, 231

Richardson, E. H., 138
Richmond, E. A., 11, 177, 199
Ricksecker, L. E., 33, 37
Ripley, L. B., 24, 38, 71, 74, 95, 102, 120, 145, 147, 169, 177, 178, 199, 202, 203, 231
Risler, J., 70, 74
Roark, R. C., 112, 138, 229–231
Robertson, C., 263
Roth, L. M., 231
Roubaud, E., 29, 38, 114, 138
Rudolfs, W., 110, 121, 131, 132, 138, 200, 231
Rumbold, C. T., 129, 138

S

St. George, R. A., 98, 100
Salt, G., 38, 263
Sang, J. H., 57, 58, 74
Sawamoto, T., 262
Schmidt, A., 166, 169
Schreiber, A. F., 200
Schulz, P., 205, 231
Schwarz, E., 40, 74
Scott, L. B., 171, 181, 186, 200
Searls, E. M., 223, 231
Secrest, J. P., 149, 168, 174, 198
Seitz, A., 263
Sen, S. K., 11, 29, 38
Senior-White, R., 131, 138
Severin, H. C., 19, 38, 138, 248
Severin, H. H. P., 19, 38, 138, 248
Sharma, H. N., 29, 38
Sharp, W. E., 34, 38
Shaw, A. O., 221, 222, 227, 231
Shaw, F. R., 213, 231
Shaw, J. G., 96, 102
Shepard, H. H., 202, 231
Sherwin, C. P., 75, 76, 101, 104, 105, 136
Shields, S. E., 219, 228
Siegler, E. H., 68, 74, 173, 200
Sihler, H., 38
Sladden, D. E., 263
Smirnow, D. A., 205, 231
Smith, C. E., 94, 102
Smith, F. F., 252, 263
Smith, R. C., 183, 200, 227, 231
Snapp, O. I., 67, 74, 81, 149, 169
Soraci, F. A., 7
Speyer, E. R., 138, 242, 245, 248
Spuler, A., 194, 200
Stabler, E. M., 248
Stabler, R. M., 200
Stäger, R., 207, 231
Stahmann, M. A., 74

Staniland, L. N., 192, 200
Starr, D. F., 102, 114, 139
Steiner, L. F., 91, 92, 102, 194, 200
Stone, W. E., 135, 197
Storch, H., 154, 168
Sweetman, H. L., 206, 225, 231, 263
Swingle, H. S., 67, 74, 81, 149, 169

T

Takahashi, R., 263
Tattersfield, F., 212, 227
Taylor, N. B., 247
Taylor, N. W., 248
Terherne, R. C., 11
Thomson, R. C. M., 11
Thomssen, E. G., 231
Thorpe, W. H., 27, 38, 99, 102, 148, 170, 253, 255–257, 263
Tinbergen, N., 110, 139
Totze, R., 11, 13, 38
Townsend, C. H. T., 27, 38
Trägardh, I., 3, 11
Traube, J., 235, 248
Travis, B. V., 23, 25, 38
Trim, A. R. H., 244, 248
Trouvelot, B., 40, 70, 74
Tschirch, A., 43, 44, 46, 74
Turnbull, W. A., 214, 232

U

Ullyett, G. C., 27, 38

V

Valentine, J. M., 21, 38, 264
Van Antwerpen, F. J., 225, 232
Vanderplank, F. L., 109, 139
Van Dyke, E. C., 34, 38
Van Leeuwen, E. R., 91, 92, 102, 177, 178, 195, 200
Vanskaya, R. A., 118, 139
Van Someren, V. G. L., 74
Varley, G. C., 264
Veillon, R., 29, 38, 114, 138
Venard, C. E., 132, 136
Verrall, A. F., 123, 139
Verschaffelt, E., 3, 11, 40, 51, 56, 74
Volkonsky, M., 232
Von Loesecke, H., 95, 102

W

Wain, R. L., 205, 232
Walker, J. C., 73, 74
Wallace, P., 232
Walton, C. L., 192, 200

Walton, R. R., 176, 200
Wardle, R. A., 11, 28, 36, 203, 232, 261, 264
Warnke, G., 111, 139, 160, 170
Waterhouse, G. A., 258, 264
Webb, J. E., 248
Weber, B., 101
Weber, H., 139, 232
Wegener, M., 205, 232
Wehmer, C., 74, 264
Weis, I., 166, 170
Weiss, H. B., 7, 8, 11, 264
Wesson, L. G., 110, 139
Wheeler, W. M., 124, 139
Whitcomb, W. D., 225, 232
Whitehead, F. E., 176, 200
Whitley, F. H., 173, 198
Whitten, R. R., 98, 102, 224, 232
Whittington, F. B., 8, 11, 186, 193, 199, 200
Wieting, J. O. G., 19, 38, 117, 121, 139, 153, 160, 163–165, 170
Wigglesworth, V. B., 11, 235, 244, 248, 264
Wilford, B. H., 98, 102
Williams, C. M., 35, 38, 180, 200
Williams, O. L., 223, 232
Williston, S. W., 264
Wilson, H. F., 139, 183, 200
Wilson, J. L., 230
Wingo, C. W., 174, 200
Wirth, W., 160, 170
Woerdeman, H., 164, 169
Woglum, R. S., 213, 232
Wojtusiak, R. J., 38
Wolcott, G. N., 224, 232
Woodside, A. M., 91, 100, 172, 173, 178, 180, 198
Worthley, H. N., 19, 38, 92, 102, 173, 178, 184, 200
Wright, E., 123, 139
Wright, J. H., 230

Y

Yasiro, H., 120, 139
Yates, F., 221, 227
Yetter, W. P., 2, 11, 91, 92, 102
Yothers, M. A., 91, 92, 102, 194, 196, 200, 211, 232
Yusope, M., 115, 136

Z

von Zehmen, H., 232
Zellner, J., 125, 129, 130, 139
Zwaardemaker, H., 141, 170

Subject Index

A

Absorption, in relation to repellency, 208
Acacia Cavenia, 62
 Farnesiana, 62
Acanthopsyche junodi, 71, 202
Acceptance, related to attractancy, 242
Acerates, 254
Acetaldehyde, 66, 67, 78, 83, 90, 91
Acetic acid, 29, 98, 110, 114, 117, 219, 234, 237
  as a codling moth attractant, 90, 91
  in fungi, 127, 129
  as an oriental fruit moth attractant, 93
  production of, by bacteria, 77, 83, 84
  responses of *Drosophila* to, 3, 95
  threshold of response to, 162
Acetic ether, 95
Acetobacter aceti, 77
 gluconicum, 77
 melanogenum, 77
 orleanse, 77
 xylinoides, 77
 xylinum, 77
Acetone, 77, 83, 84, 90, 93
Acetyl cyclohexane, 95
  methyl carbinol, 77, 83, 84, 89, 95
Achroea, 21
  interspecific copulation, 23
Acids, as codling moth attractants, 84ff.
  comparative stimulative efficiencies of, 236ff.
  formation of, in nature, 16, 79, 80
  as taste-substances, 236ff.
Aconitum Napellus, 62
Action potentials, 157
Activity coefficients, correlated with thresholds, 234
Acuity, olfactory, 163
Adsorption, by repellents, 208
Aedes aegypti, 215, 217
Aerobacillus polymyxa, 82ff.
Aerobacter aerogenes, 82ff.
 oxytocum, 83ff.
Aesculin, 57
Agaricales, 124
Agarics, as insect food, 123
Agromyzidae, 26
Agrotis, 124
Alanine, 110, 119, 181
Albumin, blood, 108, 213
 egg, 119, 181
Alcohols, 16, 77, 78

Alcohols, as codling moth attractants, 84ff.
  formation of, in nature, 16, 80
  as fruit fly attractants, 95ff.
  as taste substances, 234
Aldehydes, as codling moth attractants, 90
  as fruit fly attractants, 95ff.
Aldoses, 77, 79, 80
Algae, insects feeding on, 130
Alkyl isothiocyanates, 52
  sulfides, 29, 53, 113
Allium, 53, 260
 azureum, 53
 cepa, 53
 Porrum, 53
 ursinum, 54
Allyl alcohol, 25, 53
  amine, 25
  bromide, 53
  isothiocyanate, formula of, 52
  propyl disulfide, formula of, 54
  sulfide, 29, 113, 134
   formula of, 51, 53
  thiocyanate, formula of, 51
  thiourea, 225
Alnus, 45
Aluminum sulfate, 212
Alyssum saxatile, 51
Amanita muscaria, 125, 129, 130
 pantherina, 129
Amanitol, 126
Ambrosia beetles, feeding habits of, 123
Amelanchier vulgaris, 55
Amines, 106, 107, 113ff.
  as blowfly attractants, 113
  occurrence in fungi, 130
  occurrence in myxomycetes, 130
  as wireworm attractants, 253
Amino acids, 83, 107, 108, 181
  occurrence in fungi, 130
  in myxomycetes, 130
Ammonia, 116ff., 134, 208
  as a boll weevil attractant, 65
  as a fruit fly attractant, 181
  as a housefly attractant, 29
  as a mosquito attractant, 110
  occurrence in nature, 106–108, 116ff.
  threshold of response to, 160, 163, 165
Ammonium carbonate, 70, 107, 110, 112, 128, 134
  as a fruit fly attractant, 118, 190
  as a housefly attractant, 29
  hydroxide, 107, 110

271

Ammonium nitrate, 118
  phosphate, 118, 181
  sulfate, 118, 120, 181
  sulfide, 213
Amygdalase, 48
Amygdalin, 48, 55, 66
Amyl acetate, 68, 93, 98, 181, 213
  alcohol, 3, 78, 84, 95, 117
  benzoate, 242
  caproate, 86
  formate, 86
  mercaptan, 112
  salicylate, 223, 242
  valerate, 86
Anabasine, 211
Anabasis aphylla, 211
Anartia, 39
  jatrophae jamaicensis, 40
Anastellorhina augur, 28
Anastrepha, attractants for, 94ff., 119, 181
  ludens, 96, 119
  mombinpraeoptans, 120
  striata, 119, 120
  suspensa, 120
Andropogon Nardus, 62, 211
Anemotaxis, 9, 14
Anethole, 59, 60, 62, 64, 100, 177
  as a codling moth attractant, 68, 91–93, 178
Anethum graveolens, 59, 60
Angelica atropurpurea, 59
Anions, stimulation by, 239
Anise, oil of, 181, 216
Anise ketone, 59
Anisic acid, 62, 64, 68, 93
  aldehyde, 59, 62, 68, 93, 183
  formula of, 64
Anisol, 25
Anisotoma cinnamomea, 125
Anopheles funestus, 110
  gambiae, 1, 110
  karwari, 131
  maculatus, 131
Antagonism, 216
Anthomyiidae, 53, 115, 117
Anthonomus grandis, 65, 147
Anthriscus Cerefolium, 60
Ants, fungus gardens of, 123
  taste thresholds of, 166, 167
Aphidius, 26
Aphis bakeri, 256
Apiin, 57
Apium graveolens, 59, 60
Apple maggot fly, attractants for, 119, 120, 181
  races of, 120

Apple maggot fly, traps for, 187, 190
Arabia, 1
Arabinose, 78, 235
Arabis alpina, 51
Arabobiose, 75
Arachidic acid, 116
Arctias selene, 21
  orientation of males, 23
Argynnis, 39
Arilus cristatus, 35
Aristolochia, odors of, 130
Aristolochia-feeders, 258
Aromatic acids, 16, 85
  as codling moth attractants, 86ff.
Arracac xanthorrhiza, 59
Arsenic, effect on fermentation, 91
Artemisia biennis, 63
  caudata, 62, 63
  dracunculoides, 61
  Dracunculus, 61, 63
Asclepias, 41, 254
  chemistry, of, 71
Ascomycetes, as insect food, 123
Aseroë, 124
Ash, as a repellent, 202
Asparagine, 110, 128, 253
Aspartic acid, 110, 119, 253
Aspergillus elegans, 78
  fumaricus, 78
  nidulans, 78
  niger, 78
  ochraceus, 78
  violaceofuscus, 78
  Wentii, 78
Assembling, 21
Atalantia ceylonica, 259
Attractance, relation of chemical structure to, 242
Attractants, for carpophagous insects, 66, 80
  categories of, 4
  chemical nature of, 12
  defined, 4
  distinguished from repellents, 162
  of doubtful significance, 33
  food-type, defined, 18
  history of, 2ff.
  to increase feeding, 173
  larval, 173
  man-made, 34
  mixtures of, 177
  odorous, role of, 18
  ovipositional-type, 26ff.
    defined, 18
  seasonal variation, 93, 99
  specificity of, 19

Attractants, uses of, 171ff.
  by Romans, 2
  for xylophagous insects, 67, 80
Attractive odors, source of, 42
Attractiveness, summation of, 110
Aubrietia deltoidea, 51
Autographa, baits for, 94
  basigera, 94
  biloba, 94
  brassicae, 94
  oo, 94
  ou, 94
  oxygramma, 94
  verruca, 94
Azalea, 45

B

Bacteria, fermentation by, 77–80
Bacterium acidi lactici, 77
  amvlobacter, 77
  asiaticus mobilis, 77
  cellulose dissolvens, 77
  cloacae, 77
  coli, 77
    communis, 77
  lactis aerogenes, 77
  mesentericus, 77
  subtilis, 77
  tartaricus, 77
  xylinum, 77
Bait traps, to compare attractants, 140
Baits, analysis of, 81
  for codling moths, 82ff., 92, 178
  defined, 175
  for eye gnats, 183
  fermented by bacteria, 82, 83
  for flies, 2
  for flying insects, 176
  for fruit flies, 181
  for grasshoppers, 2, 176
  for Japanese beetles, 176ff.
  for nonflying insects, 176
  for peach moths, 2, 181
  for tobacco hornworm, 181
Baldwin and Lloyd Morgan principle of organic selection, 257
Banana, fermentation products of, 95
Barbarea vulgaris, 51
Bark beetles, feeding-type attractants for, 96
  repellents for, 224
Barosma serratifolia, 62, 63
  venustum, 62, 63
Basilarchia, 255
Bassa tribe, use of repellents by, 1

Bay laurel, oil of, 216
Bedouins, use of repellents by, 1
Benzaldehyde, 25, 48, 78, 160, 183, 242
  distribution in plant kingdom, 54
  formation of, 48
Benzene hydrocarbons, 50
Benzoates, effect on fermentation, 91
Benzoic acid, 85, 89, 107, 110
Benzol, 160
  chloride, 222
Benzyl benzoate, 69, 91, 181, 219
  chloride, 222
  cinnamate, 87
  ether, 91, 94
  isothiocyanate, formula of, 54
  valerate, 94
Bergamot, oil of, 91, 216
Betaine, 127
Betula alba, 62
Betulaceae, oils in, 62
Bismuth carbonate, 234
Blatta orientalis, 183
Blood, as a blowfly attractant, 175
  composition of, 108
  as a fruit fly attractant, 119, 181
Blood albumin, as a spreader, 213
Blood plasma, human, composition of, 107
Blowflies, attractants for, 28, 133, 174
  exposed to antagonistic stimuli, 157–159
  olfactometer for, 144, 146
  orientation to odors, 28
  oviposition of, 28, 112, 133, 174
  repellents for, 220ff.
  sheep, attractants for, 112, 113, 133
  tarsal reception by, 234ff.
Boiling point, correlated with rejection thresholds, 234
  and olfactory stimulation, 244
Bolboceras gallicus, 125
Boletus edulis, 129
  luridus, 129
  viscidus, 129
Bombycidae, scent organs in, 21
Bombyx mori, 21
Bordeaux mixture, 1, 213
Borneol, 49, 211
Boswellia serrata, 62
Brachinus, 205
Brassica campestris, 52
  napus, 52
  nigra, 52
  oleracea, 51
Breeding seasons, determination of peak, 173
Bromatia, 122, 123

Bromethyl acetate, beta, 75
Bromostyrol, 82, 91
Bruchus quadrimaculatus, 203
Buddleia variabilis, 62
Bunias orientalis, 51, 53
Buprestidae, 80, 96, 100
    attractants for, 33, 100
Burseraceae, oils in, 62
Butyl acetate, 67, 86, 246
    alcohol, 53, 77, 78, 83, 84, 245, 246
    benzoate, 87
    butyrate, 86, 246
    citrate, 87
    disulfide, 91
    lactate, 87
    dl-malate, 87
    oxalate, 87
    propionate, 86, 246
    sulfide, 91
    tartrate, 87
    α-toluate, 87
    valerate, 86, 246
Butylene glycol, 77, 83
Butyric acid, 13, 29, 95, 113, 115, 205, 237
    as a codling moth attractant, 84, 89
    in fungi, 125, 127, 129
    production by bacteria, 77
    as a sex attractant, 25

C

Cabbage looper moths, attractants for, 94
Cadaverine, 107
    formula of, 114
Caenurgia erechtea, 183
Calcium carbonate, 29, 113
    hydroxide, 213
    malate, 67
    oxalate, 128
Callidium antennatum, 67
Calliphora erythrocephala, 115, 157
    orientation of, to meat, 28
    taste thresholds of, 166, 167
Calliphoridae, oviposition attractants for, 115
Camphor, 43, 207
    structural formula of, 49
Camphoric acid, 98
Canarium luzonicum, 62
Cannabis indica, as a repellent, 1
Cantharellus cibarius, 129
Capparidaceae, attractants from, 52
Capparis spinosa, 52
Capric acid, 115
Caproic acid, 25, 77, 115, 127, 129, 177
Caprylic acid, 115

Capsella Bursa Pastoris, 51
Carabidae, at baits, 100
Carabus spp., 205
Carbitol acetate, 222
Carbohydrates, classification of, 79
Carbolic acid, 214
Carbon cycle, 17
Carbon dioxide, 77, 89, 90, 91, 106, 107, 117ff., 132, 164
    as a mosquito attractant, 110, 121, 132, 182
    as a repellent, 121
Carbonic acid, 77
Cardamine hirsuta, 51
Caricaceae, attractants from, 52
Carpocapsa pomonella, 6, 80, 81ff.
Carpophilus dimidiatus, 115
Carrot rust fly, repellents for, 225
Carum Carvi, 59, 60
Carvacrol, 48, 50
Carvone, 59, 60, 62
    structural formula of, 50
Caryophyllene, 24
Casein, 119, 181, 213
Cassia, oil of, 216
Catacola, 40
Cations, comparative stimulative efficiencies of, 240
Cecidomyidae, 26
Cedarwood, oil of, 216
Cellobiose, 75
Cellulose, 16, 77
    origin of essential oil from, 47
Cephalosporium luteum, 123
    pallidum, 123
Cerambycidae, 80, 96, 100
Ceratia similis, 202
Ceratitis, attractants for, 94ff., 118, 178, 181
    capitata, differential response of sexes to attractants, 19, 34
Ceratostomella ips, 123
    montium, 123
    piceaperda, 123
    pseudotsugae, 123
Ceroplatidae, 124
Cetyl alcohol, 126
Chaetodacus, attractants for, 94ff., 118, 181
    dorsalis, 118, 194
    jarvisi, 118
    tryoni, 118, 213
Chara contraria, 131
    foetida, 131
    fragilis, 131
    hispida, 131
    intermedia, 131
    as a repellent, 131

Cheiranthus Cheiri, 62
Chemical structure, relation of attractance to, 242
relation of repellence to, 242
Chemoreception, physiological factors affecting, 255
Chemoreceptors, tarsal, 234ff.
Chenopodiaceae, oils in, 62
Chenopodium vulvaria, 49, 66, 114
Chiggers, repellents for, 219
Chlorides, comparative stimulative efficiencies of, 240
Chlorine substitution, effect on threshold, 237
Chloronaphthalene, 222
Chlorophyll, origin of essential oils from, 47
Cholesterol, 108, 110
Choline, 126, 127
Chromodoris, 236
Chrysanthemum cinerariaefolium, 211
Chrysochus auratus, 7
Chrysomyia albiceps, 28
bezziana, 28
varipes, 28
Chrysopidae, bait preferences of, 98
Chrysotoxine, 127
Cicuta bulbifera, 59
maculata, 58
Cineol, 48, 49
Cinnamic aldehyde, 43, 183
Cinnamomum zeylanicum, 43
Cicuta virosa, 59
Cistus, 47
Citral, 21, 29, 43, 64, 68, 91, 93, 100
Citrene, 98
Citric acid, 78, 79, 89, 98, 129, 234
Citripestis sagittiferella, 29
oviposition by, 64
Citronella, attraction to *Dacus*, 3, 13
oil of, 68, 91, 92, 211, 216, 222
Citronellal, 81, 176, 211
Citronellol, 50, 176, 211
Citrus Aurantium, 44
Medica, 61
vulgaris, 43
Citrus borer, oviposition by, 29
Citrus thrips, sprays for, 174
Clasterosporium, 78
Clathrus, 124
Clausena Anisum-olens, 63
Claviceps purpurea, 124
Clavin, 127
Clensel, as an attractant, 118
Cleome spinosa, 52
Clove, oil of, 68, 91, 92, 95, 192, 207, 216, 222

Clysia ambiguella, 21
response to female-baited traps, 23
Coccophagus caridei, 255
Coccus fagi, 254
Cochlearia Armoracia, 51
officinalis, 52
Cochliomyia americana, 28, 112, 222
Cockroaches, repellents for, 225
Codling moth, attractants for, 81ff., 115, 242
baits for, 82, 92, 178
control by trapping, 196
emergence dates, 172, 173
oviposition by, 80
ovipositional-type attractants for, 68
repellents for, 211
response of, to color, 6
sex ratio, 194
sprays for larvae of, 173, 174
traps for, 193, 194, 196
Collagen, 108
Collembola, 110
Collinsonia anisata, 62
Color, relation of trap efficiency to, 8, 192
response to, 6ff.
Colorado potato beetle, attractants for, 70
repellents for, 213
Common chemical sense, 207
Compositae, oils in, 62, 63
Concentration, effect of, 117
of essential oils, 57
optimum, 161ff.
regulation of, 179
related to repellency, 207
reversal, 163, 167
threshold, 161
Conditioning, to chemicals, 256
to determine threshold, 156, 167
to odors, 21, 29
Conidia, as attractants, 124
Coniferin, 57
Conium maculatum, 58
Conotrachelus nenuphar, 66, 80
reactions in an olfactometer, 67
Contarinia pyrivora, 192
Control, by trapping, 195–197
Copper, compounds of, as repellents, 221
in magnesium oxychloride cement, 226
pentachlorophenate, 224
resinate, 213
stearate, 213
Copra beetles, attractants for, 115
Coprinus lagopus, 124
Coprophagous insects, attractants for, 106, 111, 114, 117ff.
Corcyra cephalonica, 183

Coriandrol, 59, 60, 64
Coriandrum sativum, 60
Corn, attractants from, 69
Corn earworm moth, attractants for, 69
  olfactometer for, 149
  oviposition by, 29
  sprays for, 174
Cosmos, 60
Cotinus nitida, attractants for, 115
Cotoneaster tomentosa, 55
Cotton, attractants from, 65, 114, 147
Crambe cordifolia, 51
Crataegus Oxycantha, 55
  Pyracantha, 55
Creatinine, 108
Creep, related to duration of repellency, 216
Creophilus, food of, 19
  villosus, 111
Creosote, 213
Cresol, 222
Cresylic acid, 222
Crotonaldehyde, 222
Crotonyl isothiocyanate, formula of, 52
Cruciferae, chemistry of, 51ff., 62
Cryolite, 202
Cryptotermes brevis, 224
Culex, attraction of, 109
  pipiens, oviposition by, 29
Culicoides, 211, 219
Curculionidae, at baits, 100
Curcuma longa, 62
Cybocephalus, 124
Cyclohexane, 160
Cyclohexylamine, 114
Cydonia japonica, 55
  vulgaris, 55
Cyllene pictus, 67
Cymene, structural formula of, 50
Cyrillia, 27
Cystine, 108, 109, 119, 181
  bacterial decomposition of, 113

D

Dacnusa, 26
Dacus, attractants for, 94ff., 118, 181
  cucurbitae, 120, 213
  differential response of sexes to attractants, 19, 24
  diversus, 19
    response to methyl eugenol, 24
  ferrugineus, response to eugenol, 24
  niger, 118
  oleae, 118, 120
  response to citronella, 3, 13

Dacus, zonatus, 19
  responses to iso-eugenol, 24
Danais, 166
  plexippus, 41, 70, 254
Daucus Carota, 59
Daviesia latifolia, 75
Decomposition products, 18
Decyl alcohol, 223
Deguelin, 213
Delphinin, 57
Dendrobium superbum, 116
Deodorizers, 208
Derris, 211, 213
Desmia funeralis, 194
Dextrinose, 75
Diacetyl, 95
Diallyl sulfide, formula of, 54
Diamines, 107, 113
Diatoms, as food for insects, 130
Dibasic acids, 85
  as codling moth attractants, 86ff.
Dicranura vinula, 205
Dictamnus Fraxinella, 60, 62, 63, 259
Dictyophora, 124
Diethyl ketone, 90
Diethylene glycol monobutylether acetate, 223
Digilanide, 57
Dihydrocarveol acetate, 24
Dihydroxyacetone, 77
Dihydroxybenzoic acid, 78
Diluents, 179
Dimethyl phthalate, 216–219
Dimethyl sulfide, 54
Dimethylamine, 66
  formula of, 113
Dimethylaniline, 222
Dioxane, 95
Diphenyl ether, 94
  methane, 24, 95
  oxide, 91
Diploidization, flies as agents of, 124
Dippel's oil, 213
Diptera, parasitic, attractants for, 28
Disaccharides, 75, 79
Disonycha quinquevittata, 7
Distribution coefficients, correlated with thresholds, 234
Disulfides, 59
  formation of, from cystine, 113
Diversinervus smithi, 249
Dixylylguanidine, 225
Dodecyl alcohol, 223
Drosophila, attractants for, 94ff., 177
  differential response of sexes to attractants, 19, 95

Drosophila, guttifera, 256
  melanogaster, response of, to air, 9, 14
  preimaginal conditioning of, 256
  repellence of dry air to, 15
  response of, to acetic acid, 3, 95, 114,
    162
    to amyl alcohol, 3
    to ethyl alcohol, 3, 95, 161
    to lactic acid, 3
  traps for, 189
Dulcitol, 79, 235
Dust, particle size, 202
  as a repellent, 9, 201, 202
Dytiscus, 168

E

Echinophora spinosa, 59
Elastin, 108
Elymus arenarius, 124
Embidobia, 26
Empleurum serrulatum, 61
Empoasca abrupta, 252
  bifurcata, 252
  erigeron, 252
  fabae, 213, 252
  filamenta, 252
  maligna, 252
Emulsifiers, 180
Emulsin, 48
Emulsions, field tests of attractive, 88
Endomyces bispora, 123
Endomychidae, 124
Entomiasis, accidental, 28
  habitual, 28
Entomiasis-producers, oviposition of, 28,
  112
Entomophagous insects, attractants for,
  106
Environmental factors affecting feeding
  habits, 99
Enzymes, in fungi, 126
  glucoside hydrolysis by, 48, 52
Ephestia, 21, 255
  cautella, 183
    odor glands of, 22
  elutella, 183
  figulilella, 194
  kühniella, 183
    odor glands of, 22
  relation of, to its parasites, 27
Epilachna varivestis, 213
Ergosterine, 126, 127
Ergotinine, 127
Ergotoxine, 127
Erysimum Perofskianum, 51

Erythritol, 79, 235
Essential oils, 39, 42ff., 206
  chemical constituents of, 48ff.
  classification of, 49ff.
  defined, 42
  dispersion of, 47
  effect of climate and soil on, 48
  location in plant, 42
  methods of controlling concentration
    of, 57
  physiological status of, 47
  receptacles for, in plants, 43ff.
  as repellents, 216
  resinification of, 45
  secretion of, 43ff.
  from Umbelliferae, 56
Esters, as codling moth attractants, 86ff
Ethyl acetate, 78, 86, 246
  alcohol, 3, 53, 84, 95, 97, 191, 217, 245,
    246
    as a housefly attractant, 117
    produced by bacteria and fungi, 76–
      78, 83
    threshold of response to, 160, 161, 164
  amine, 25
  anisate, 25
  benzoate, 25, 67, 87
  butyrate, 25, 47, 86, 88, 246
  cinnamate, 87
  citrate, 87
  lactate, 87, 88
  malonate, 87
  mercaptan, 112, 134
  oxalate, 87
  oxyhydrate, 81, 86, 93
  pelargonate, 25
  phenyl acetate, 88
  propionate, 86, 88, 91, 246
  succinate, 87, 88
  tartrate, 87, 88
  α-toluate, 87, 88
  p-toluidine, 91
  valerate, 86, 246
Ethylidene aniline, 91
Eucalyptus, oil of, 216
Eugenol, 24, 25, 43, 50, 93, 177, 194,
  222
Eupeleteria magnicornis, 27
Euphoria inda, 99
Euploea asela, morphology of scent organs
  of, 22
European corn borer, attractants for, 69
Evaporation, retardation of, 179
Excitation, compound field of, 157
Eye gnats, baits for, 183
  traps for, 188, 192

F

Fabrics, impregnation of, 218
Fats, 108
  classified, 103
Fatty acids, 107, 114ff., 126ff.
  classified, 105
  in fungi, 129
  in myxomycetes, 130
  unsaturated, 115, 116
Feces, composition of, 107
Feeding habits, categories of, 30
  inheritance of, 255
  mutations affecting, 260
Feeding preferences, 249
Female odor hypothesis, 24
Fennel, oil of, 68, 91, 93, 94, 192
Fermentation, changes produced by addition of arsenic and sulfites, 91
  cyclic production of compounds in, 96
  formula for, 76
  products of, 18, 117
    attractive to xylophagous insects, 96
    by bacteria and fungi, 77–80
    of bananas, 3
  repellents formed by, 70, 208
  specificity of odors of, 98
Feronia elephantum, 259
  limonia, 259
Ferula, 260
Fibrinogen, 108
Firebrat, repellents for, 225
Flea beetles, repellents for, 213
Flies, olfactometer for, 148, 150
  testing repellents for, 220
    whole-cow method, 221
Fluoranthrene, 224
Fluorene, 224
Foeniculum piperitum, 50
  vulgare, 59, 60
Food cycle, 17
Food plant selection, 39ff., 206, 249ff.
  affected by repellents, 206
  factors influencing, 40, 131
Food plants, effect of, on trapping, 193
Food-type attractants, 30ff.
  classification of, 30
Forcipomyia, 34
Formaldehyde, 93
Formic acid, 77, 83, 84, 127, 129, 205, 237
Fortisia, 27
Fortunella crassifolia, 259
Fraxinus, 45
Fructose, 68, 83, 92, 234
Fruit flies, 13, 19, 24, 119ff.
  attractants for, 94ff., 118ff., 134, 181, 243

Fruit flies, control by trapping, 197
  olfactometer for, 145, 147
  sex ratio, 194
Fruit fly, cherry, 206
  Jarvis, bait for, 118
  mediterranean, bait for, 118
  Mexican, attractants for, 120, 192
  olive, bait for, 118, 181
  Queensland, bait for, 118
    repellents for, 213
  solanum, bait for, 118
Fucus, 35
Fuligo septica, 127, 129
Fumaric acid, 78, 79, 83, 126, 129, 238
Fundamental animal products, 18
Fundamental plant products, 18
Fungi, acids in, 129
  amines in, 130
  amino acids in, 130
  attractants in, 124, 128
  blue stain, 123
  chemistry of, 125ff.
  fatty acids in, 129
  fermentation by, 77–80
  as insect food, 123ff.
  in relation to termite feeding, 224
  relations between insects and, 31, 123
  secretions of, as attractants, 124
  wood-staining, 123
Fungin, 126, 127
Fungus gardens, 122, 123
Furfural, 67, 68, 93, 242
Fusarium lini, 78
Fuscosclerotic acid, 127

G

Galactose, 78
Galleria, 21
  interspecific copulation by, 23
  mellorella, 27
Gallic acid, 67
Gases, as attractants, 90, 121
Gasteromycetes, 124
Gastroidea viridula, 69
Gelatin, 119, 181
Genetic basis of attraction, 162
Genetic basis of repellency, 162
Gentianose, 76
Gentiobiose, 75
Geotrupes, 111
  stercorius, 160
  sylvaticus, 160
  vernalis, 160
Geraniol, 43, 48, 71, 176, 211
  structural formula of, 50

Geranyl acetate, 91, 216
  formate, 81
Glands, of Compositae, 45
  of Labiatae, 45, 46
  of Pelargoniae, 45
  repugnatorial, 205
Gliodin, 110
Globulin, blood, 108
Glossina palpalis, 211
Gluconic acid, 78, 80
Glucose, 48, 76–78, 83, 92, 100, 108, 126, 234
Glucosides, as attractants, 56
  hydrolysis of, 48, 52
  naturally occurring, 16, 57, 75
  origin of essential oils from, 47
Glucoxylose, 75
Glutamic acid, 108, 119
Glutamine, 253
Glycerol, 68, 77–79, 91, 128, 173, 234
  diborate, 222
Glycine, 83, 108, 119, 181
Glycocoll, 109
Glycogen, 128
Gnaphalium obtusifolium, 71
Gonia capitata, 27
Gossypium, 254
Gradient fields, 13
Gramineae, oils in, 62
Granulobacter pectinovorum, 77
Grapholitha molesta, 6, 80, 92, 181
Guanine, 128
Gustatory repellents, 206, 222
Gypsy moth, assembling of, 23
  attraction of males to female abdomens, 20
  sex attractants for, 114
  sex scent of, 24

H

Habrobracon, 255
  juglandis, 21, 110, 160
  orientation of, to host, 27
  preimaginal conditioning of, 256
  response of males to female odor, 22
Haematobia irritans, 211
Heat of convection, 4
Heliothis obsoleta, 69
Helminthosporium geniculatum, 78
Hemerocampa leucostigma, 41, 146
Hemoglobin, 108
Hemp, as a repellent, 1
Heptaldehyde, 25
Hesperis matronalis, 51
Hessian fly, 205

Heterofermentative bacteria, 82ff.
Hexachlorophenol, 224
Hexose, 77, 79
Hexyl alcohol, 217
Hibiscus syriacus, 254
Hippelates, trap for, 188, 192
Hippuric acid, 107
Honey bee, bait preferences of, 98
  repellents for, 213
  search for food by, 21
  taste thresholds of, 166, 167
Hopkins host principle, 257
Host odors, vertebrate, 28
Host preferences, physiological factors influencing, 255
Host selection, genetic basis of, 255
Houseflies, olfactometer for, 152, 153
  olfactory thresholds of, 160, 164, 165
  oviposition by, 29, 118
  repellents for, 223
  traps for, 187, 191
Humidity, 14
  as an attractant, 14
    for mosquitoes, 110, 132
  as a factor affecting threshold, 164
  reactions, 14, 15, 132, 133
  receptors, 15
  as a repellent, 14
  response to, by Agriotes, 15
    by Drosophila, 15
    by Lucilia, 15
    by Nemobius, 15
    by Porcellio, 14
    by Ptinus, 15
    by sheep blowflies, 133
  at surface of transpiring leaves, 57, 58
Hydnocystis arenaria, 125
Hydnum repandum, 129
Hydrochloric acid, 42, 205, 208
Hydrocotyle americana, 59
  umbellata, 59
Hydrocyanic acid, distribution in plant kingdom, 54
  formation of, 48
Hydrogen, 77, 90
  as an attractant, 121
Hydrogen ion, concentration gradient in beets, 250
  regulating feeding, 236, 250ff.
  stimulating effect of, 236
Hydrogen sulfide, 29, 106, 107, 133
Hydrolysis, of glucosides, 48, 52
Hydroxy acids, 16, 85
  as codling moth attractants, 86ff.
  production of, 116
Hydroxybenzoic acid, 83

Hydroxyl substitution, effect on threshold, 237
Hylemyia antique, 53
    reactions of, to mustard oils, 53
Hylurgopinus rufipes, 224
    attractants for, 98
Hymenoptera, oviposition of parasitic, 26
Hyssopus, 45
Hysterangium, 124

I

Iditol, 79
Illicium anisatum, 62
    verum, 62
Impregnated fabrics, 218
Indalone, 216, 219
India, 1
Indol-acetic acid, 107
Indol-ethylamine, 107
Indole, 29, 47, 106, 107, 110, 111ff., 134, 160
    as an attractant, 111, 134
    formula of, 111
Inheritance, of host preferences, 255
Injection of chemicals into trees, 97
Insecticides, as repellents, 207
Iodine, 205
Ionic mobilities, 240
Ionone, 48
Ips fasciatus, 97
    grandicollis, 96
    pini, 96
Irone, 48
Isoamyl acetate, 86
    alcohol, 25, 84, 245, 246
    amine, 25
    benzoate, 69, 87, 181
    butyrate, 86, 246
    mercaptan, 25
    oxalate, 87
    propionate, 86, 246
    salicylate, 69, 181
    tartrate, 87
    valerate, 246
Isobutyl acetate, 86
    alcohol, 84
    amine, 113
    benzoate, 87
    butyrate, 86
    lactate, 87
    methyl carbinol, 89
    phenyl acetate, 81, 91
    propionate, 86
    α-toluate, 87
    valerate, 86

Isocellobiose, 75
Isophenylbutyl acetate, 93
Isopropyl acetate, 86, 246
    alcohol, 83, 84, 86, 245, 246
    benzoate, 87
    lactate, 87
    meta-cresol, 48
    ortho-cresol, 48
    oxalate, 87
    propionate, 86
Isopulegyl formate, 24
Isoquinoline, 222
Isothiocyanate of secondary butyl alcohol, formula of, 52
Isovaleric acid, 77

J

Japanese beetle, baits for, 176
    oviposition by, 29, 225
    plants immune to, 212
    population reduction, 172, 197
    repellents for, 212, 225
    response of, to color, 7, 8
    traps for, 179ff.
Jassidae, 99
    attraction to baits, 99
        to maize, 174
June beetle, repellents for, 225
    response to sex attractants, 23, 114

K

Kelp flies, attractant for, 35, 180
Keratin, 108, 113
Kerosene, 24, 91
Kerria, 55
Ketones, as codling moth attractants, 90
Ketoses, 77, 79, 80
Kineses, 4
Klinokinesis, 26
Kru tribe, use of Malvaceae by, 1

L

Labiatae, glands in, 45
    oils in, 62
Lactarius piperatus, 129
    vellereus, 129
Lactates, 77, 108
Lactic acid, 3, 25, 77, 83–85, 95, 108, 110, 115, 127, 129, 238
Lactobacillus pentoaceticus, 77
Lactones, 70
Lactose, 75, 92

Lasiocampidae, scent organs in, 21
Lathyrophthalmus arvorum, attractants for, 114
Lathyrus latifolius, 52
   sylvestris, 52, 69
Lauraceae, oils in, 62
Lauric acid, 115
Lauronitrile, 223
Laurus nobilis, 62
Lavandula vera, 48
Lavender, oil of, 216
Leaf area, measurement of, 210
Leaf sandwich tests, 42
Leafhoppers, regulation of feeding, 236, 250ff.
Lecithin, 113, 126, 128
Leguminosae, oils in, 62
Lepidoptera, olfactometer for larvae, 154–156
Leptinotarsa decimlineata, 70
Leptocera, attractants for, 118
Leucine, 78, 126, 127
Levidulinose, 76
Liberia, 1
Libocedrus decurrens, 62
Lice, reaction of, to radiant heat, 5
Light, orientation to, 5–8, 157–159
Limanthaceae, attractants from, 52
Lime, 212, 213
Limonene, 68, 91
Limonius californicus, 25
   canus, 25
Linalool, 43, 48, 67, 93, 181
Linalyl acetate, 67, 95, 216
Linolenic acid, 116
Linolic acid, 93, 116, 181
Linseed oil, 93
   soap of, 116
Liparis dispar, 21
   monacha, 6, 206
Lipids, attractants from, 106
   chemical changes in nature, 105
   classified, 103
   decomposition products of, 105
Lippia, 40
   adoensis, 62
Lithobius, 27
Litsea odorifera, 61
   sericea, 62
Locusts, African migratory, 14
Loganiaceae, oils in, 62
Lonchocarpus, 211, 212
Lophanthus anisatus, 62
   rugosa, 62
Lucilia cuprina, oviposition of, 112
   sericata, 28, 112, 115, 134, 157

Lycaena hypophlaeas, 70, 254
Lycoperdon Bovista, 130
Lysigenous oil storage space, 43

M

Mace, oil of, 91
Magnesium sulfate, 206
Magnolia Kobus, 62
Magnoliaceae, oils in, 62
Malacosoma americana, 160
   food plants of, 55
Maleic acid, 238
Malic acid, 68, 78, 83, 89, 98, 126, 129, 173, 206, 234, 238, 253
   esters of, 82
Malonic acid, 85, 89
Maltose, 75, 92
Malvaceae, 1
Mandelonitrile glucoside, 48
Manninotriose, 76
Mannitol, 77, 79, 126, 235
Mannonic acid, 78
Mannose, 78, 127
Masking, of odors, 208
Mealy bugs, control by banding, 202
Meat, change of attractiveness with age, 112
Mediterranean fruit fly, bait for, 118
Megarhyssa atrata, 26
   lunata, 21
Melaleuca minor, 46
Melanophila acuminata, 34
   consputa, 33
   ignicola, 34
   longipes, 33
   notata, 34
Melasoma populi, 205
Melezitose, 76
Melia azedarach, 212
Melibiose, 75
Melinus minutiflora, 212
Meliphora grisella, 27
Melon fly, attractants for, 116
   oviposition by, 120
   repellents for, 213
Membracidae, attraction to baits, 99
Mentha aquatica, 62
   crispa, 219
   longifolia, 62
   piperata, 46, 219
   spicata, 62
   velutina, 62
   verticillata var. strabala, 62
Menthol, 46, 49, 158, 159
Menyanthes trifoliata, 53

Mercaptans, 106, 111ff.
  as blowfly attractants, 134
  formation of, from cystine, 113
Mercuric chloride, 221, 222
Meroden, 225
Mespilus germanica, 55
Metachlorophenol, 219, 222
Metaponasthru, 27
Methallyl disulfide, 222
Methane, 29, 77, 107
  as an attractant, 121
Methyl acetate, 86
  alcohol, 67, 84
  allephenol, 68, 93
  allyl thiocyanate, 112
  amine, 25, 66, 127
    formula of, 113
  benzoate, 87, 242
  butyrate, 86
  chavicol, 59, 60, 62, 63
    formula of, 64
  cinnamate, 68, 82, 87, 93, 94
  ethyl ketone, 90
  formate, 67
  lactate, 86, 87
  malonate, 87
  mercaptan, 107
    formula of, 111
  naphthalene, 91
  nonylketone, 48, 61, 64
  oxalate, 86, 87
  phenyl acetate, 25
  propionate, 86
  salicylate, 25, 222, 242
  α-toluate, 87
Mexican bean beetle, repellents for, 213
Mexican boll weevil, 254
  attractants for, 65
  olfactometer for, 147
  olfactory thresholds of, 163
Mexican fruit fly, attractants for, 120, 192
Microplectron fuscipenne, 27
Mite, cheese, aggregation of, in concentration gradient, 14
  itch, 1
Modalities, of odors, 241
  of taste, 240
Molecular moment, correlated with thresholds, 234
Molecular surface, correlated with thresholds, 234
Molecular weight, related to odor intensity, 245
Monilia, 78
  brunnea, 123

Monilia, candida, 123
Monobasic acids, 77
Monochloronaphthalene, 224
Moringaceae, attractants from, 52
Mosquitoes, host selection by, 132
  larval food plants, 131
  testing repellents for, 209, 214ff.
Mothproofing, 225
Mucor mucedo, 78
  racemosus, 78
Musca domestica, 160, 223
Muscaridine, 126
Muscarine, 126
Muscina, attractants for, 118
Mustard oils, 51
  distribution of, in plant kingdom, 52
Mutations, affecting feeding habits, 260
Mutinus, 124
Mycetophagous insects, 121ff.
Mycetophilidae, 124
Mycose, 127
Myosin, 128
Myrcene, 43
Myristic acid, 115, 116
Myrosin, 52
Myrrhis odorata, 60
Myrtaceae, oils in, 62
Myxomycetes, acids in, 129
  amines in, 130
  amino acids in, 130
  fatty acids in, 129

N

Naphthalene, 207, 213, 224, 225
Naphthols, alpha and beta, 214
Naphthylamine, alpha, 222
Narcissus fly, repellents for, 225
Natal fruit fly, responses of, 95
Necrobia rufipes, 115
Necrophagous insects, attractants for, 106, 111, 114
Necrophorus americanus, 111, 160
  orbicollis, 111, 160
Nematodes, root selection by, 252
Nemeritis canescens, host selection by, 27
Nemobius griseus, 15
  palustris, 15
  response of, to humidity, 15
Nerve, action potentials of, 157
Nicotiana tabacum, 211
Nicotine, 211
  sulfate, 91, 92, 178, 213
Nitrobenzene, 192
Nitrogen, cycle, 17
  oxides of, 205

Noctuidae, bait preferences of, 98
Notodonta concinnula, 205
Nymphalis antiopa, 71
  cardui, 71
Nyssa sylvatica, 259

O

Obscurants, 208
Ocimum Basilicum, 62
  gratissimum, 62
  sanctum, 62
Odors, caste, 25
  classification of, 241
  conditioning to, 256
  defined, 12, 233
  as a factor influencing food plant
    choice, 40
  hive, 25
  intensity of, 244
  measurement of, 140
  mixtures of, 177
  modalities of, 241
  source in nature, 15
  sting, 25
  thresholds of response to, 160
  vertebrate, 109
  wax, 25
Oecanthus niveus, 203
Oidium, 78
Oil, cavity in *Ruta graveolens*, 47
  storage space, formation of, in plant,
    43ff.
    lysigenous, 43
    schizogenous, 43
    schizolysigenous, 43
Oleic acid, 77, 93, 98, 110, 116, 125, 127,
  129, 181, 222
Olfactometer, 140ff.
  Barrows', 142–144
  defined, 141
  for lepidopterous larvae, 154–156
  McIndoo's, 40, 143–147
  requirements for ideal, 149
  for testing blowflies, 144, 146
    boll weevils, 147
    corn earworm moths, 149
    flies, 148, 150
    fruit flies, 145, 147
    houseflies, 152, 153
    wireworms, 146, 147
  tests of codling moth baits in, 85ff.
  Venturi type, 151ff.
  Y-tube type, 142ff.
  Zwaardemaker's, 141
Olfactory organs, 233

Olfactory prism, 241
  threshold, 156ff., 160
    methods of determining, 156
Oligophagy, chemical basis of, 254
Oncodes incultus, 99
Orchelimum glaberrimum, 29
Orchidaceae, oils in, 62
Organic bases, 51, 64
Organic compounds, synthesis by green
  plants, 16
Oriental fruit moth, attractants for, 92, 93
  baits for, 181
  ovipositional-type attractants for, 68, 92
  repellents for, 213
  response of, to color, 6
  sex ratio, 194
Orientation, of *Calliphora* to meat, 28
  directed, 4
  to host, 26
  to light, 157–159
  secondary, 4
  to sex odors, 23
  undirected, 4
Ortalidae, bait preferences of, 98
Orthochlorophenol, 219, 222
Orthodichlorobenzene, 224
Orthokinesis, 14
Ortho-tolylthiourea, 225
Osmorrhiza longistylis, 59, 60
Oviposition, of blowflies, 28
  factors governing, 26, 133
  inhibited by repellents, 205, 225
  by phytophagous insects, 29
Ovipositional-type attractants, for blood-
  sucking insects, 29
  for fruit flies, 120
  for oriental fruit and codling moths, 68
Oxalic acid, 33, 69, 78, 79, 85, 89, 126,
  129, 206, 234
Oxybenzyl isothiocyanate, formula of, 54
Oxygen, 90
  effect on stimulation, 242
Oxypolis filiformis, 59
Oxyria digyna, 69

P

Packaged goods, protection of, against
  insects, 225
Palmetto weevil, 35, 180
Palmitaldehyde, 94
Palmitic acid, 116, 125, 127, 129
Papilio, 58ff., 258
  ajax, choice of food plants by, 58ff., 254
  oviposition by, 64
  response to odors, 59

Papilio, alexanor, 258, 260
  andraemon, 259
  bairdi, 258
  constantinus, 258
  cresphontes, 259
  demoleus, 258
  evolution of feeding habits of, 258
  hospiton, 258, 260
  indra, 258
  machaon, 258, 259
  ophidecaphalus, 258
  paeon, 258
  zelicaon, 258
Paracholesterin, 127
Paracresol, 106, 107
Para-dichlorobenzene, 222
Paraffin, 213, 222
Para-methoxybenzoic acid, formula of, 64
Para-methoxypropenylbenzene, 64
Para-oxyallylbenzene, 64
Para-oxyphenylacetic acid, 107
Para-oxyphenylpropionic acid, 107
Parasites, attraction to baits, 99
  host selection by, 27, 132, 249
Para-xylidine, 222
Pastinaca sativa, 59
Peach, attractants from, 66
Peach borer, feeding habits of, 66
  oviposition of, 80
Pear midge, 192
Pectins, 16, 77
Pelargonin, 57
Pelea madagascarica, 62
Penicillium griseo-fulvum, 78
  chrysogenum, 78
  digitatum, 78
  oxalicum, 78
  purpurogenum, 78
  spiculisporum, 78
Pennyroyal, oil of, 216, 222
Pentachlorophenol, 224
Pentose, 77, 79
Pepper maggot fly, sprays for, 174
Pepsin, 128
Peptone, 77, 119
Perilitus, 26
Periophorus padi, 55
Persea carolinensis, 259
  gratissima, 62
Petroleum oils, attraction of high boiling
  fractions, 34
Phallalis, 124
Phallus, 124
  impudicus, 129
Phenanthrene, 224
Phenolthiazine, 213

Phenols, 16, 50, 67, 107, 108, 213, 221,
  222
Phenyl methyl carbinol, 89
Phenylacetaldehyde, 94
Phenylacetic acid, 83
Phenylacetylcarbinol, 78
Phenylalanine, 110
Phenylcyclohexanol, 216
Phenylethyl alcohol, 47
  amine, 107
  butyrate, 177
  carbinol, 88, 89
  cinnamate, 87
  isothiocyanate, formula of, 52, 54
  phenyl acetate, 87
Phenylglycine, 67
Phenylhydrazone, 67
Phenylthiourea, 225
Philanthus triangulatus, 110
Phlebotomus papataci, 130
Phorbia, attractants for, 118
Phormia regina, 115, 222
Phosphatides, 108
Phosphorus, 108
Photosynthesis, 15
Phycophagous insects, attractants for, 130
Phyllophaga anxia, 225
  lanceolata, 23
Phytomonas vascularum, 124
Phytophagous insects, attractants for, 3
  host preferences of, 39
  repellents for, 212ff.
Pieridae, change in feeding habits of, 260
Pieris brassicae, 25, 51
  rapae, 51, 134, 160
Pilocarpus Jaborandi, 61
Pimenta acris, 62
  leaf oil, 177, 179
Pimpinella Anisum, 60
Pimpla ruficollis, 99, 255
Pinaceae, oils in, 62
Pine, repellents from, 221, 224
Pine tar oil, 91, 92, 177, 178, 216, 221, 222
Pinene, 98, 222
Pinites succinifer, 62
Pinus contorta, 62
  Jeffreyi, 62
  palustrus, 62, 221
  Sabiniana, 62
  sylvestris, 99, 255
Piper aduncum, 259
  Betel, 63
  mollicomum, 259
  peltatum, 62, 259
  umbellatum, 259
Piperaceae, oils in, 62, 63

Piperonal, 25, 68
Pirus, 55
Pitch, 1, 181
Plant attractants, 3, 42
Plantago major, 42
Plants, extraction of repellents from, 210
  resistant varieties, 202ff.
Plastin, 128
Platypezidae, 124
Platypus compositus, 123
Platysamia cecropia, 240, 254
Plodia, 21, 183
  interspecific copulation, 23
Plum curculio, feeding habits of, 66
  oviposition of, 80
Poanes hobomok, 40
Pollenia stygia, 28
Polychrosis botrana, 21
  response to female-baited traps, 23
Polygnotus, 26
Polygonaceae, attractants in, 69
Polyporus officinalis, 126, 130
Polysaccharides, 16, 79
Polysaccum pisocarpium, 129
Pomace flies, traps for, 186, 189, 190
Pontomyia natans, 130
Populin, 57
Populus pyramidalis, 259
Porcellio, 14
Porosagrotis orthogonia, 27
Potassium hydroxide, 113
Potato leafhopper, repellents for, 213
Preimaginal conditioning, 256
Primverose, 75
Propionaldehyde, 88, 90
Propionic acid, 77, 80, 84, 90, 114, 125,
  129, 237
Propyl acetate, 86, 88, 93
  alcohol, 53, 78, 84, 86
  amine, 49, 66
  butyrate, 86, 246
  lactate, 87
  oxalate, 87
  propionate, 86, 246
  valerate, 246
Proteins, 77, 83, 118, 126, 127
  attractants from, 106, 119, 181
  classified, 103
  as fruit fly attractants, 119, 181
  hydrolysis of, 104, 119, 181
Protoparce sexta, 68, 181
Protubera, 124
Prunase, 48
Prunasin, 57
Prunus avium, 55
  cerasus, 55

Prunus Laurocerasus, 53, 55
  lusitanica, 63
  Padus, 55
  Persica, 55
  virginiana, 55
Psalliota, 129
Pseucedanum graveolens, 60
Ptelea trifoleata, 61
Pterandrus, 94, 118, 178
  rosa, 95
Pterocyclon fasciatum, 123
  mali, 123
Pteroselinum sativum, 58
Pubescence, plant, as a repellent, 8, 203
Puccinia graminis, 124
  helianthi, 124
Putrefaction, odors of, 111
Putrescine, 66, 107
  formula of, 114
Pycnia, as attractants, 124
Pyrameis, 166, 167
Pyrausta nubilalis, 69
Pyrene, 224
Pyrethrins, 211
Pyrethrum, 211, 216, 219, 221
Pyridine, 96, 213
Pyruvic acid, 237

Q

Quercus ilicifolia, 42
Quinic acid, 77

R

Radiant heat, 4, 5
Raffinose, 76
Ranunculaceae, oils in, 62
Recognition odors, 25
Rejection, related to repellency, 242
Repellent Index, 217
Repellents, action of, 201
  affecting food plant choice, 206
  aquatic plants as, 131
  categories of, 4
  cosmetic acceptability, 214
  criteria for ideal, 214, 220
  defined, 4, 201
  distinguished from attractants, 162
  evaluation of, 157, 159, 174, 208ff.
  extraction from plants, 210
  formed by fermentation, 70, 208
  gustatory, 206, 222
  history of, 1ff.
  influencing reproductive behavior, 205
  inhibiting oviposition, 205, 225
  of insect origin, 205

Repellents, as insecticides, 207
  mode of action, 226
  naturally occurring chemical, 205–207
  olfactory, 207ff.
  physical, 201
  relation of chemical structure to, 242
  screening of, 208ff., 219ff., 223
  specificity of, 207
  synthetic, 212ff.
  for termites, 224
  use of, by Bassa people, 1
    by Greeks, 2
    by Kru people, 1
Reproduction, behavior of, influenced by
    repellents, 205
Reseda, 51
  alba, 52
  lutea, 52
  luteola, 52
  odorata, 52
  virgata, 52
Resedaceae, essential oils of, 52
Resin, 39, 127, 128, 206
Resinogenous layer, 44, 45
Resistant factors of plants classified, 204
Resistant varieties of plants, 202ff.
Rhagoletis, attractants for, 94ff., 119, 181
  juglandis, 119
  pomonella, 119, 190
Rhamninose, 76
Rhamnose, 235
Rheum, 69
Rhizopertha dominica, 29
Rhizopogon, 124
Rhizopus nigricans, 78
Rhododendron, 45
Rhyacionia buoliana, 99, 255
Rhynchophorus cruentatus, 35, 180
Rhyphus, 117
Rhyssa, 26
Rioxa musae, 118
Robinose, 76
Root-dwelling insects, repellents for, 224
Root-feeders, host selection by, 252
Rosaceae, attractants in, 54ff., 63
Rose geranium, oil of, 216
Rotenone, formula of, 211
Roubieva multifida, 62
Rumex Acetosella, 70, 254
  hydrolapathum, 58
  scutatus, 69
Russula foetida, 124
Ruta chalepensis, 61
  graveolens, 47, 61
  montana, 61
  patavina, 61

Rutaceae, chemistry of, 61ff., 259
Rutaceae-feeders, 258
Rutgers #612, 216
Rutinose, 75

S

Safrol, 43, 68, 91, 93, 181, 222
Salicin, 57
Salicylaldehyde, 67, 205
Salicylic acid, 93
Saligenin, 25
Salts, comparative stimulative efficiencies
    of, 240
  as taste substances, 240
Salvia, 45
  officinalis, 53
Sampling of populations, 172, 195
Santalyl acetate, 216
Saprophagous insects, attractants for, 106
Sarkine, 128
Sassafras officinale, 46
  oil of, 91, 92, 216
  variifolium, 42
Saturniidae, scent organs in, 21
Scatella subquattata, 130
Scelio, 26
Scent glands, morphology of, 22
Scent organs, 21
Schizogenous oil storage space, 43
Schizolysigenous oil storage space, 43
Scitamineae, oils in, 62
Sclererythrin, 127
Sclerojodin, 127
Scleropristallin, 127
Sclerotic acid, 127
Scleroxanthin, 127
Scolytidae, 80, 96, 100, 123
  association with fungi, 123
Scolytus, 98
  multistriatus, 224
    attractants for, 98
  ventralis, 123
Scototaxis, 6
Screen tests, 41
Seasonal variation of attractants, 93, 99
Secalinaminosulfo acid, 127
Secondary butyl isothiocyanate, formula
    of, 54
Secretion of essential oils, 43ff.
Sedanolid, 59, 60
Sedanonic anhydride, formula of, 64
Seed treatment, 213
Selinum, 60
Seseli, 260
Sesquiterpene, 207

Sex, ratio in traps, 19, 92, 95, 119, 194
Sex attractants, 20, 21ff., 175
  for adult wireworms, 25
Sex scents, chemistry of, 24, 25
  dispersion of, 23
Sheep, attractants from, 112, 174
Silvanus advena, 115
Silverfish, repellents for, 225
Sinapis arvensis, 51
Sinigrin, 52
Sirex, 26
Sisymbrium officinale, 51
  strictissimum, 51
Sium circutaefolium, 58
Skatole, 14, 29, 106, 107, 110, 111ff., 160
  as an attractant, 111, 134
Skin, human, composition of, 108
Smoke, as an attractant, 33
  as a repellent, 1
Soap, 93, 98, 108, 113, 194
  as a melon fly attractant, 116
Sodium arsenite, 91, 118
  benzoate, 89, 91, 92
  bicarbonate, 206
  bisulfate, 67
  chlorate, 98, 191
  chloride, 29, 42, 91, 107, 206
  citrate, 29
  fluosilicate, 213
  hydroxide, 113, 119, 181
  hypochlorite, 91
  oleate, 68, 91, 93, 98
  silicofluoride, 225
  sulfide, 29, 113
  sulfite, 91
  tartrate, 29
Solanaceae, 70
Solidago, 61
  odora, 63
  rugosa, 63
Solubility, related to chemical stimulation, 246
Sorbitol, 68, 79, 173, 234, 235
Sorbus americana, 55
  Aucuparia, 55
Spiraea, 55
Spirogyra, as food for insects, 130, 131
Spores, secretions of, as attractants, 124
Sprays, attractants added to, 193
  timing of, 173
Standard free energies, correlated with stimulation, 234
Stapelia flavirostris, 112
Staphylinidae, at baits, 100
Starch, 16, 77
  origin of essential oil from, 47

Stearic acid, 110, 116, 127, 129
Steriphoma paradoxum, 52
Sterols, 108
Sticky bands, 191, 202
Stimuli, categories of, 4
  chemical, 4
    classes of, 17
  hygro-, 14
  mechano-, 8, 201
  photo-, 5
  physical, 4
  thermo-, 4
  token, 19ff.
  use of opposed, to determine thresholds, 157–159
Stomoxys calcitrans, 109, 115
  repellents for, 211
Strophanthobiose, 75
Strumigenys pergandei, hunting by, 110
Succinic acid, 78, 83, 85, 86, 93, 129, 238
Sucrose, 42, 68, 75, 83, 234
Sugar, as an adjuvant to poison sprays, 173, 174
  alcohols, occurrence in nature, 79, 80
  cane, gumming of, 124
  fermentable, 109
Sulfides, 111ff., 133
  addition of, to blowfly baits, 113
  as blowfly attractants, 113, 133
Sulfites, effect on fermentation, 91
Sulfur, 219, 225
Sweat, as affecting repellency, 215
  apocrine, 109
  as an attractant, 109
  composition of, 109
  eccrine, composition of, 108
Synathedon exitiosa, 67, 80
  pictipes, 67
Synergic Index, 217
Synergism, 216
Syrphidae, bait preferences of, 98

T

Tabanidae, bait preferences of, 98
Tachinidae, 27
  host relations of, 27
Tagetes minima, 212
  minuta, 62
Tannic acid, 206, 234
Tannin, 126
  origin of essential oil from, 47
Tar, 224
Tarsal stimulation, factors limiting, 238
Tartaric acid, 77, 89, 93, 98, 126, 129, 206, 234, 238

Taste, definition of, 12, 233
  modalities of, 240
  organs of, 233
  as a repellent, 206, 222
  response of carpophagous larvae to, 68,
    173, 174
  source in nature, 15
  thresholds of, 166–168
Taste-substances, categories of, 234
Taxes, 4
Taxodium distichum, 62
Temperature, as an attractant, 4, 109,
    132
  gradient, 5
  optimum, 5
  as a repellent, 4
Tent caterpillars, repellents for, 213
Tephrosia, 211, 212
Termites, attractants for, 224
  food preferences of, 223
  fungus gardens of, 122, 123
  repellents for, 207, 223
Terpene, 128
  origin of essential oils from, 47
  structural formula of, 49
Terpinyl acetate, 24, 93, 181, 216
Tetralol, 216
Tetramethyldiamine, 66
Tetrasaccharides, 79
Thanite, 221, 222
Thelaira, 27
Thiocyanates, as repellents, 211
Thiourea, 225
Threshold, absolute, 156
  conditions affecting, 164
  olfactory, 156, 160
  of response, 162
  taste, 166–168
Threshold concentration, defined, 161
Thrips, attractants for, 183
  traps for, 188
Thrixion, 27
Thyme, oil of, 216
Thymol, 48, 50
Thymus citriodorus var. *montanus*, 62
  vulgaris, 48
Ticks, 1
  in concentration gradient, 13
  repellents for, 219
Tilletia Tritici, 124
Tissue choice, by phytophagous insects,
    249
Tobacco hornworm, 68
  attractants for, 69, 181
  trap for, 186, 188
Toluic acid, alpha, 85, 86

Toxin, 126
Tradescantia, 40
Tragacanth, 85
Trapping, effect of adjacent plants on, 193
  efficiency for control, 195–197
  to sample populations, 172, 195
Traps, for apple maggot flies, 187, 190
  for bark beetles, 191
  color of, 8, 192
  defined, 183
  electrocuting, 191
  for eye gnats, 188, 192
  for fruit flies, 189, 190
  for grapevine moths, 2
  for houseflies, 187, 191
  for Japanese beetles, 179ff.
  for male gypsy moths, 175
  metal, wooden, and glass, 183ff.
  New Jersey mosquito, 182, 187
  for pomace flies, 186, 189, 190
  position and location of, 192, 193
  for reducing populations, 172
  for thrips, 188
  for timing broods, 173
  for tobacco hornworm, 186, 188
Traube series, 235
Trehalose, 75
Tremex columba, 26
Trichlorethylene, 35, 180
Trichogramma, 26
  evanescens, 205
  host selection by, 27
Trichosporium symbioticum, 123
Tridecylonitrile, 223
Trimethylamine, 47, 49, 65, 66, 113, 124,
    126, 127, 163
  formula of, 113
Triphasia, 259
Trisaccharides, 76, 79
Triticum durum, 205
  vulgare, 205
Tropaeolaceae, attractants from, 52
Tropaeolum, 51, 54
  majus, 52
  peregrinum, 52
Trypetidae, 80, 99
  attractants for, 94ff., 119ff.
  feeding habits of, 120
  ovipositional-type attractants for, 120
Trypsin, 110
Tsetse fly, attractants for, 109
  host selection by, 109
  repellents for, 211
Turanose, 75
Turpentine, 206, 213, 219
Tyrolychus casei, 14, 111

Tyrosin, 110, 119
Tyrothrix tenius, 77

U

Umbelliferae, chemistry of, 56ff.
  distribution of essential oils in, 60
  essential oils of, 94, 259
  odors of, 59
Umbelliferae-feeders, 258
Undecenyl alcohol, 223
Undecyl alcohol, 223
Undecylonitrile, 223
Urea, 107–109
  derivatives of, as repellents, 225
Uric acid, 107–109
Urine, human, composition of, 107
  as a repellent, 1
Ustilaginales, 124

V

Valeraldehyde, 90
Valerianic acid, 29
Valeric acid, 25, 84, 91, 92, 110, 113, 115, 178, 237
Vanilla, as a fruit fly attractant, 118
Vanilla planifolia, 62
Vanillyl alcohol, 25
Vapor pressure, correlated with thresholds, 234
Verbascum Thapsus, 42
Verbenaceae, oils in, 62
Vernine, 127
Vicianose, 75
Vinyl sulfide, formula of, 54
Vinylite binders, 219
Viola, 45
Viscosity, of repellents, 216
Vitellin, 128

W

Walnut husk fly, attractants for, 119, 181
  oviposition by, 120

Water, as a repellent, 9, 201
Water content of plants, 203
Wave length, response to, 6ff.
Wax moth, 27
Whitewash, 213
Wintergreen, oil of, 216, 222
Wireworms, olfactometer for, 146, 147
  root selection by, 253
  sex attractants for, 25
Wohlfartia magnifica, 28
Wood, repellent factors in, 223
Wood-boring insects, repellents for, 223
Woodlice, aggregation of, in moist air, 14
Wood naphtha, 221
Wood-staining fungi, 123
Wool, attractants from, 113
  bacterial decomposition of, 113

X

Xanthine, 126, 128
Xyleborus affinis, 123
  pecanis, 123
Xylosandrus germanus, 97
Xylose, 78, 235

Y

Yeast, 78, 89, 95, 97, 119, 181

Z

Zanthoxylum ailanthoides, 61
  ajuda, 259
  americanum, 61, 259
  Clava-Herculis, 259
  coriaceum, 259
  schinifolium, 61
  senegalense, 61
  simulans, 61
Zinc sulfate, 213

Date Due

MAR 7	Levin	
MAY 1 4	Mc Wright	
JUN 8 '65	Wirzoanaty	
MAR 1 5 '66	Glassman	
MAY 5	Burgess	
NOV 1 0 '60	Thomas	
JAN 19 70	Panagopulas	
MAY 25 '70	Tuscori	
OT 16 75		
APR 8 2 '77	Miller	
SEP 2 7 1979	Schmidt	
DEC 9 1980	Gruber	
JAN 26 1987	Co	JAN 2 0 REC'D
ⓐ		